South
Atlantic
Quarterly
Winter 2008
Volume 107
Number 1

Visit Duke University Press Journals at www.dukeupress.edu/journals.

Subscriptions. Direct all orders to Duke University Press, Journals Customer Service, 905 W. Main St., Suite 18B, Durham, NC 27701. Annual subscription rates: print-plus-electronic institutions, $178; print-only institutions, $159; e-only institutions, $158; individuals, $35; students, $21. For information on subscriptions to the e-Duke Scholarly Collection through HighWire Press, see www.dukeupress.edu/edukecollection. Print subscriptions: add $13 postage and 6% GST for Canada; add $17 postage outside the U.S. and Canada. Back volumes (institutions): $159. Single issues: institutions, $40; individuals, $14. For more information, contact Duke University Press Journals at 888-651-0122 (toll-free in the U.S. and Canada) or 919-688-5134; subscriptions@dukeupress.edu.

Permissions. Photocopies for course or research use that are supplied to the end user at no cost may be made without explicit permission or fee. Photocopies that are provided to the end user for a fee may not be made without payment of permission fees to Duke University Press. Address requests for permission to republish copyrighted material to Permissions Coordinator, permissions@dukeupress.edu.

Advertisements. Direct inquiries about advertising to Journals Advertising Coordinator, journals_advertising@dukeupress.edu.

Distribution. The journal is distributed by Ubiquity Distributors, 607 DeGraw St., Brooklyn, NY 11217; phone: 718-875-5491; fax: 718-875-8047.

The *South Atlantic Quarterly* is indexed in *Academic Abstracts FullTEXT Elite, Academic Abstracts FullTEXT Ultra, Academic Research Library, Academic Search Elite, Academic Search Premier, America: History and Life, Art Index Retrospective, 1929–1984, Arts and Humanities Citation Index, Corporate ResourceNet, Current Abstracts, Current Contents/Arts and Humanities, Discovery, Expanded Academic ASAP, Historical Abstracts, Humanities Abstracts, Humanities and Social Sciences Index Retrospective, 1907–1984, Humanities Full Text, Humanities Index, Humanities Index Retrospective, 1907–1984, Humanities International Complete, Humanities International Index, International Bibliography of Periodical Literature, Literary Reference Center, MasterFILE Elite, MasterFILE Premier, MasterFILE Select, MLA Bibliography, News and Magazines, OmniFile Full Text V, OmniFile Full Text, Mega Edition, Research Library, Social Sciences Index Retrospective, 1907–1984,* and *Student Resource Center College with Expanded Academic ASAP.*

The *South Atlantic Quarterly* is published, at $178 for (print-plus-electronic) institutions and $35 for individuals, by Duke University Press, 905 W. Main St., Suite 18B, Durham, NC 27701. Periodicals postage paid at Durham, NC, and additional mailing offices. Postmaster: Send address changes to *South Atlantic Quarterly,* Box 90660, Duke University Press, Durham, NC 27708-0660.

© 2007 by Duke University Press

ISSN 0038-2876

The Agamben Effect
SPECIAL ISSUE EDITOR: ALISON ROSS

Introduction 1
ALISON ROSS

Playing with Law: Agamben and Derrida on Postjuridical Justice 15
CATHERINE MILLS

The Saturday of Messianic Time (Agamben and Badiou on the Apostle Paul) 37
ELEANOR KAUFMAN

The Inversion of Exceptionality: Foucault, Agamben, and "Reproductive Rights" 55
PENELOPE DEUTSCHER

Particularity and Exceptions: On Jews and Animals 71
ANDREW BENJAMIN

Bare Life on Strike: Notes on the Biopolitics of Race and Gender 89
EWA PŁONOWSKA ZIAREK

Agamben: Aesthetics, Potentiality, and Life 107
CLAIRE COLEBROOK

Except for Law: Raymond Chandler, James Ellroy, and the Politics of Exception 121
LEE SPINKS

Suspended Animation: Thinking and Animality
in Neurocultural Selfhood   145
ADRIAN MACKENZIE

Witnessing the Inhuman:
Agamben or Merleau-Ponty   165
JEAN-PHILIPPE DERANTY

After Humanism: Agamben and
Heidegger   187
KRZYSZTOF ZIAREK

Notes on Contributors   211

Alison Ross

Introduction

In his 1995 book *Homo Sacer: Sovereign Power and Bare Life*, Giorgio Agamben presents his philosophical project as the attempt of a genealogy of the occidental conception of "life." More than this, his project aspires to a new conception of the determination of the political in modernity, which would bring to a fundamental level of articulation the incomplete reflections on biopolitics in the work of Hannah Arendt and Michel Foucault. His articulation of the biopolitical paradigm calls on a far more extensive range of considerations than either of his precursors had envisaged. The reason for this must be sought in his definition of *life* as a politically determined concept. In this vein, he cites the definitions of life in medical practice alongside those categories explicitly located in the domain of political contestation, such as the refugee, as already political determinations.

As made clear in his criticisms of Foucault in the introduction to *Homo Sacer*, in Agamben's view biopolitics has its origins in the thinking of the political in the West and is not fully comprehensible when it is understood either as a distinctive feature of the modern period or as modernity's prevailing set of institutional operations. His view that thinking today needs to return to the path of "first philosophy" or "ontology" defines the orientation that shapes this perspective. His preference for "ontology" indicates the extent of his departure from the approach to biopolitics taken by Foucault, which was skeptical of the resources of philosophical discourse to grapple with historical problems and especially of its capacity to instruct political action. In contrast, it is clear that from the first formulations of his project Agamben had anticipated its significance as far more than a diagnosis of the fundamental tendencies in the politics of occidental modernity; it is, in fact, more like the promise of "an answer to the bloody mystification of a new global order."[1] Indeed, despite the historical purview of his project, it is Agamben's position that the significant tendencies of occidental biopolitics reach an unprecedented degree of intensification today, and he accordingly calls on particular features of the contemporary political scene to give his project force and urgency.

This project has since been developed in successive publications, including *Means without End: Notes on Politics* (1996), *Remnants of Auschwitz: The Witness and the Archive* (1998), *The Open: Man and Animal* (2002), and *State of Exception* (2003). One of the central problems in this collection of works is the definition of "bare life" or "naked life" (*nuda vita*).[2] In *Homo Sacer* it is the political determination of bare life that occupies his attention. Like Foucault, Agamben thinks that modernity is characterized by an increasingly more radical tendency to take control of "life." For Agamben, the significance of this tendency can be measured against the distinction that classical political philosophy (Aristotle) maintains between *zoē*, bare or naked life, and *bios*, the life constituted in the *polis*. In what is now a characteristic gesture, he wishes to foreground the dire consequences of the *zoē-bios* distinction for the life that is naked or bare, the latter understood in terms of its complete exposure to sovereign action. Like Arendt's concept of a human being completely stripped of his or her rights, Agamben's notion of bare life wishes to put in view the disposable status of such life when it is utterly exposed to political calculation. In other words, bare life is human life that is completely exhausted in its status as the correlate of sovereign action. He argues that the two separate spheres of naked life and politics defined by Aristotle have become fused in the modern period. Politics is now increasingly defined as an era of biopolitics, in which power is exercised as rule over life. The frequent references Agamben makes to "aporia" and "zones of indistinction" are ways of marking this feature of the modern era, which he finds in the stateless refugee in the camp, the shifting definitions of life in medical practice, and the bare life of the prisoner in the concentration camp.

In *Remnants of Auschwitz*, Agamben takes his genealogical project into the Nazi death camps to examine how the camps produce bare life. Although this work presents itself as an account of the ethical aporia and imperatives of witnessing in which the capacity to accommodate the extremity of the death camps would provide the test for any elaboration of ethical precepts, the analysis has a political dimension, given Agamben's view, articulated in a number of places, that it is the camp and not the state that is "the fundamental biopolitical paradigm of the West" (*HS*, 181). It is in *The Open*, however, that he attempts to explicitly render as a subject of analysis the biopolitical discourse of the Occident. This discourse, which Agamben phrases as the "anthropological machine of humanism," aims at the separation of the human from the animal in humans themselves and leads toward the production of (available) bare life.[3] The significance of this work is not just its account of the production of bare life but the way it raises the critical question of the stakes and effects of Agamben's own project: namely, whether knowledge of the functioning of the anthropological machine may aid in stopping it. The same question is pursued from a different angle in *State of Exception*, in which Agamben states that he wants to lift "the veil covering this ambiguous zone . . . between law and the living being" in order to "answer the question that never ceases to reverberate in the history of Western politics: what does it mean to act politically?"[4]

Like the question of whether knowledge of its functioning could stop the anthropological machine, the political benefits of his analysis of the machinic functioning of the law are presented in terms of the promise of the future deactivation of the law as a "disused object" to be "played with" and "studied" (*SE*, 64). Aside from this type of utopian gesture that prophesies the destinal suspension of the effects of law and seems to make any action superfluous, what these studies share is their methodology. It is no exaggeration to state that Agamben's reasoning always proceeds from extreme cases or threshold states, such as the patient in intensive care, the inmate in the concentration camp, or the juridical aporia of the state of emergency.[5] These extreme examples provide far more than material for Agamben's theses on the biopolitical determination of the West. They are, in his view, both the provocation to an explanation for contemporary theory and the definitive test against which the explanatory claims of other ethical and political theories are found wanting. Hence, the following criticism of Karl Otto Apel in *Remnants of Auschwitz* makes the incontestable point that a number of theories find their limits of application in the death camps:

> Years ago, a doctrine emerged that claimed to have identified a kind of transcendental condition of ethics in the form of a principle of obligatory communication. . . . According to this curious doctrine, a speaking being cannot in any way avoid communication. Insofar as, unlike animals, they are gifted with language, human beings find themselves, so to speak, condemned to agree on the criteria of meaning and the validity of their actions. Whoever declares himself not wanting to communicate contradicts himself, for he has already communicated his will not to communicate. . . . Let us imagine for a moment that a wondrous time machine places Professor Apel inside the camp. Placing a *Muselmann* before him, we ask him to verify his ethics of communication here too. At this point, it is best, in every possible way, to turn off our time machine and not continue the experiment.[6]

In this passage Agamben identifies in the figure of the *Muselmann* a limit case able to stand as a counterproof to those attempts, typified by Apel, to hold apart the human and the animal not just on the distinguishing criteria of linguistic pragmatics but in terms of the "agreed" parameters that govern the ethical value of meaning and action. Many of the contributors to this issue take up these key themes of his work. What concerns us here, however, is the thesis regarding the explanatory scope of his own project, implied by this criticism of Apel. Leaving aside for the moment the corollary of his analyses—that the point of distinction between what is exceptional and normal is itself blurred and that it is the presumed integrity of this distinction in the functioning of legal and political institutions that conceals the contemporary normalization of the exception—Agamben wants to show how certain power relations generally thought to be only exceptional provide the key for understanding normal institutions and practices.[7] At the end of *Homo Sacer*, he acknowledges that the "lives" he lists in support of his theses on bare life at stake in biopolitics, which include the comatose patient, the neomort waiting for his organs to be transplanted, the figure of the führer, the *Muselmann* from the camps, the bandit, and Flamen Diale ("one of the greatest priests of classical Rome"), "may seem extreme, if not arbitrary" (*HS*, 182, 186). However, he insists that these lives all occupy "difficult zones of indistinction" between "law and fact, juridical rule and biological life" and that it is from their analysis that "the ways and forms of a new politics must be thought" (*HS*, 187). The mode of his approach to these lives is also telling, however, in that he wants to phrase, from the evidence of these examples, the ontological question, what is bare life? or what is a camp? and seeks to derive insights into contemporary politics from this approach (*ME*, 36). Indeed, Agamben goes so far as to suggest that the thinking of new political categories is somehow a condition for political activism today.

It helps to form an overall picture of Agamben's thought if we place it next to some of the concerns and problems that motivate the work of Foucault and Jean-Luc Nancy. The extent of Agamben's departure from Foucault's ascending model of analysis—which builds its tentative general picture by systematic reference to the analysis of how particular institutional practices operate rather than the ontological question of what they are—is instructive. Foucault's criticisms of political philosophy proceed from his contestation of their explanatory utility. In *The History of Sexuality*, he emphasizes, for instance, that the postulates of Marxist theory may be used to arrive at mutually contradictory explanations of the same phenomena and thus are unable to adequately explain anything.[8] He observes in his lectures on political philosophy that ideas from the modern tradition of contract theory, such as "legitimate power" and its conceptual partners "consent" and "the subject," block from view some important features of modern political life, especially the ways in which disciplinary practices operate. The fault, in his view, is that traditional political philosophy wants to find out what power is rather than how it functions.[9]

As a point of departure for a critical examination of Agamben's project and its political claims, we might submit his project to the dual test that Foucault sets up for political philosophy: on the one hand, it must be able to speak across "a population of dispersed events" or, in other words, not impose a reductive explanatory principle on complex phenomena; and, on the other, to its explanatory capacity to render legible a field of "dispersed events," it must add a willingness to test its hypotheses against real situations.[10] The contrast between Foucault's approach to biopolitics and Agamben's claim that the camp is the biopolitical paradigm of modernity is telling in this context. Gilles Deleuze has identified the novelty of Foucault's work not just in its break with the terminology of traditional political philosophy but in the operative distinction in his approach to power between the microanalysis of a specific institution or setting and the abstract machine or diagram, immanent to the entire social field, that this analysis suggests. Although these two aspects of Foucault's approach pose a problem of complementarity in *Discipline and Punish*, Deleuze argues that in the first volume of *The History of Sexuality* they work in concert as microdisciplines (specific institutional structures such as the arrangement of sleeping quarters in boarding schools) that are also biopolitical (immanent to the social field). Biopolitics describes the logic of the administration of life that underpins the specific disciplines but also escapes being a globalizing mode of explanation because it is arrived at by the analysis of local aims of power whose effects are not confined to a specific locale. In this context, Deleuze praises Foucault's ability to grid or map a social space from the microanalysis of its local disciplines.[11]

The status of Agamben's thought qua political philosophy may be interrogated from this perspective. In his focus on bare life, Agamben provides a set of normative terms from which it becomes possible to interrogate diverse sets of institutional practices from the perspective of the direction they take. It is precisely this normative dimension whose absence in Foucault's work critics often had cause to lament and which Agamben's work may be seen to provide. However, Agamben's claim regarding the camp as the biopolitical paradigm of modernity needs to be considered in view of its conceptual consistency as an explanatory postulate; but because it purports to form an element of political philosophy, it must be brought into relation with a testable field in which the cogency of its characterization of the modern period can be assessed. Finally, given its messianic tone, it is important to examine as well the promissory dimension that Agamben attempts to derive from his analysis of the functioning of legal instruments and categories. His approach reverses Foucault's ascending methodology and leaves us to ask what the reasoning from extreme instances tells us about the hold of Agamben's analysis on the phenomena it wishes to decode. Why does he think it necessary for a "first philosophy" (i.e., an ontology) to be the entry point into the analysis of politics? Do these distinctive elements of his approach succeed in identifying otherwise obtuse elements of political modernity? Can such insights be gleaned from or supported by other sources? How do the elements of his mode of argumentation support the nature of the claim he makes for his project's political relevance? And where do his key theses locate him in relation to other major theories of modernity?

At issue in each of the essays in this collection are the difficulties of defining the Agamben effect, especially when the analyses of his recent suite of works invest so emphatically in the motif of "inoperativity." There is the important problem to consider of whether the concepts Agamben formulates are equal to the task of providing an intellectually compelling picture of our contemporary ethical and political situation and problems as well as possible responses to these. What possibilities do his analyses foreclose in respect to praxis and the conceptualization of forms of life (Andrew Benjamin, Claire Colebrook, Jean-Philippe Deranty, Penelope Deutscher) and which ones do they open (Adrian Mackenzie, Ewa Płonowska Ziarek)? It is because Agamben's discussion of topics such as bare life is conducted through his confrontation with a number of themes and figures from the history of philosophy that many of the contributors to this issue also ask after the adequacy of his handling of these themes and figures (Deranty, Krzysztof Ziarek) or elaborate and interrogate his accounts of such themes and figures with reference to the lines of communication they open with other contemporary thinkers (Eleanor Kaufman, Catherine Mills) or genres of writing (Mackenzie, Lee Spinks).

It is possible to be clearer about the import and value of Agamben's ambitious project if we separate its diagnostic from its promissory claims. These are the two distinct registers in Agamben's recent writing, which are distinguishable at the level of the questions they pose and the types of imperatives they respond to. The diagnostic task responds to the problem of how to understand some of the compelling features of our contemporary political space and to bring to a level of reflection the experiences they shape. Here, Agamben calls on the resources of a "first philosophy" in order to identify the high stakes the category of life carries in the West and to show how the modern period represents an intensification of a tendency already present, especially in the ancients' juridical inscription of life onto the field of sovereign power. From this perspective, Agamben may be described as bringing into view specific features, or better, tendencies, of the present by situating them in relation to the previous practices and thinking from which they derive and against which they may also be distinguished. The tools he uses to conduct his analysis are also worth mentioning. In general, the outlines of his commentary are taken from a philological discussion of legal statutes and practices. (His analysis of the category of life is an obvious case in this regard, although similar etymological attention is given as well to concepts such as the sacred, the exception, and the sovereign.)

It is important to distinguish the main elements of this register from the promissory dimension of Agamben's recent writing. In this latter mode, his writing calls into question the conventional vocabulary and conception of political action and implies that his biopolitical diagnosis could instruct an adequate conception of the political for the West today. To be sure, neither this claim nor the range of activist practices he calls into question is novel. Foucault made the claim in his seminars in the early 1970s that the material of contemporary politics was impervious to the analytical hold of the tools of political philosophy and that politics had to be reconceived along the lines of a response to contemporary power relations. Moreover, he, too, was critical of the legal instruments and categories carried in human rights discourse, although unlike Agamben he tried to dissect the bases of the emerging claims to "rights," which he saw as legible from the perspective of the interstices of sovereign and disciplinary power. Neither did he believe, as Agamben often seems to suggest, that one could steer political activism by theoretical categories.[12] Thus, unlike Agamben, Foucault does not take the step of invalidating appeals to human rights as an effective tool, however limited, of political opposition and resistance.[13]

The radical perspective Agamben adopts on such topics may be understood as a consequence of the urgency he sees in our contemporary situation and the promissory implications he draws from this. Many of the signal tropes of his writing are invested with the sense that a juncture of enormous import has been reached today. The emphasis he places on the present moment sets him apart from the approach to the "history of the present" in Foucault's work and the cautious epochal approach to the era of technology in Martin Heidegger.

In terms of the themes and topics he brings to bear on his understanding of the present moment, he shares much with Nancy. Nancy's work is preoccupied with the question of what it means to live today, and he attempts to write an ontology that could accommodate this question. His starting point for the understanding of the present is Friedrich Nietzsche's diagnosis of Western nihilism, according to which "even the highest values have devalued themselves."[14]

Nancy adds to this perspective on nihilism an account of what he takes to be the signal features of the operations of contemporary capitalism in order to explain the real impetus and the urgency of changes that are occurring today, which would have no hold over life practices were they conceptual constructs elaborated solely in the philosophical tradition. In his view the operations of capital bring the groundlessness of the values of the West to a state of general awareness today. In a significant parallel to Nancy's thought, Agamben focuses on the increasing awareness of the declining semantic hold of the vocabulary of modern politics and juridical institutions (community, citizen, human rights, rule of law, etc.) but does so especially to draw attention to the "protracted eclipse" politics is undergoing in this era of "accomplished nihilism" (*ME*, i).[15] Whereas Nancy wishes to draw attention to the sources of meaning that are, on account of the prevailing operations of capitalism, for the first time seen to be made rather than given, Agamben offers something distinctly different: the need to think a new vocabulary for politics now that the categories of the citizen and the worker and the very understanding of political contestation have lost their original meanings. Hence, he refers to "pure means," "pure violence," and "the gesture" as the terms from which the new understanding of politics may be developed (*ME*, 59).

Although such suggestions remain vague and undeveloped in his work, it is clear from the settings he gives to this "new politics" that it will emerge as a result of the waning hold of what, to his mind, are the founding divisions that structure life in the West. Unlike Nancy, who sees many possible paths for a response to the waning of traditional existential regimes of meaning, none of them certain in their final outcomes or effects, Agamben focuses his attention on describing the direction he thinks our present institutional practices are taking. Indeed, the dark picture he paints of the present situation is at least partly motivated by the goal of elaborating on the theme of exhaustion, which he argues typifies our current legal institutions and practices and presages their passing. This characterization supports his contention that the founding distinction between *bios* and *zoē* is reaching an end from which a new politics will emerge. This style of analysis, which draws together the rhetoric of his diagnosis of the West with his prophetic statements regarding a new era, is coupled with the tendency to describe his analyses as "lifting veils," discovering "inner secrets," and laying "bare" the logic of the Occident (*SE*, 2, 86). In this respect he is closer to the tone of theorists of modernity such as Guy Debord and Paul Virilio, each of whom focuses on a single defining trait of the modern. In Nancy's writing, the prophetic and utopian tone of such post-Marxist theory is often criticized for its tendency to rely on a dualist style of analysis in which the present is viewed from the perspective of a more "authentic" reality or mode of human existence. Agamben's critique of the humanist mirage covering the anthropological logic of the West necessarily depends on such dualist thinking; equally, his "new politics" of "pure means" is the perspective from which the present is condemned and the link between his diagnosis of the twentieth century and his promissory tone is at its strongest.

For a writer who aspires to present a compelling picture of our contemporary situation and of possible responses to it, it is noticeable that in many ways Agamben's writing remains captivated by the classical categories and terms that thinkers like Nancy and Foucault, in radically different ways to be sure, have shown to be figments of the "past" and in which they identify obstacles to the rigorous thinking of the present. The status of art in Agamben's work is particularly telling in this respect. At crucial points Agamben calls on works of visual art to forge the link that would otherwise be missing to his view of the ending of our present darkness. He reads in Titian's painting *The Three Ages of Man* (1513–14), and especially in its depiction of the postcoital scene between nymph and shepherd, the halting of the anthropological machine and its characteristic separation of animal and human. Alternatively, he fills in the gaps of his account of the juridical aporia of our era with citations from Franz Kafka. The following passages from *The Open* and from *State of Exception*, respectively, are exemplary of these two tendencies:

> In their fulfillment the lovers learn something of each other that they should not have known—they have lost their mystery—and yet have not become any less impenetrable. But in this mutual disenchantment from their secret, they enter . . . a new and more blessed life, one that is neither animal nor human. It is not nature that is reached in their fulfillment, but rather (as symbolized by the animal that rears up the Tree of Life and of Knowledge) a higher stage beyond both nature and knowledge, beyond concealment and disconcealment. . . . As is clear from both the posture of the two lovers and the flute taken from the lips, their condition is *otium*, it is workless [*senz'opera*].[16]

> Kafka's most proper gesture consists not (as Scholem believes) in having maintained a law that no longer has any meaning, but in having shown that it ceases to be law and blurs at all points with life. (*SE*, 63)

His reference to Titian in *The Open* to present the concept of messianic time may be queried in terms of the meaning context in which a painting of a mythological scene may operate a claim of this type. Similarly, while it is true that Kafka is often appealed to as a diagnostic and prophetic writer, his suggestive literary accounts of the experience of being subjected to law cannot be cited as a source of sociological authority, nor can they stand in for an analysis of prevailing juridical codes. Although it is legitimate to ask whether elements of Agamben's analysis of the present may be used to clarify and expound on the insights of literary genres, it is important also to ask what work Agamben expects literature and art to perform for his project of characterizing the present and defining the task of forming an adequate response to it. This point is not limited to literature narrowly conceived. It is clear that Walter Benjamin's references to "pure violence" in his "Critique of Violence" require explication and cannot be called on to clarify or support Agamben's statements regarding "pure acts," "pure means," "pure violence," and "the gesture" as the idiom of the new politics.[17] Indeed, such statements call on religious faith in order to sustain and convey their meaning, and as such they demand critical analysis. One is inclined to view his handling of the artwork as the other side of a sociologically reductive conception of art: in his work, art is treated as providing real insight into social conditions and the tools to assist their passing.[18] This presumption may be valid, but in this case, as in the other cases studied by the contributors to this issue, it is important to ask for an argument that would be able to support and defend it.

Notes

1. Giorgio Agamben, *Homo Sacer: Sovereign Power and Bare Life*, trans. Daniel Heller-Roazen (1995; Stanford, CA: Stanford University Press, 1998), 12. Hereafter cited parenthetically by page number as *HS*.
2. English translations of Agamben's work render *nuda vita* as either "bare life" (as in Heller-Roazen's translation of *Homo Sacer*) or "naked life" (as in Vincenzo Binetti and Cesare Casarino's translation of *Means without End: Notes on Politics* [1996; Minneapolis: University of Minnesota Press, 2000]; hereafter cited parenthetically by page number as *ME*).
3. Giorgio Agamben, *The Open: Man and Animal*, trans. Kevin Attell (2002; Stanford, CA: Stanford University Press, 2004), 37–38.
4. Giorgio Agamben, *State of Exception*, trans. Kevin Attell (2003; Chicago: University of Chicago Press, 2005), 2. Hereafter cited parenthetically by page number as *SE*.
5. His reference to the "poetic state of emergency" in his essay "The End of the Poem," first published in Italy in the same year as *Homo Sacer* (1995), uses a tone that coincides with the suite of works that belongs to the biopolitical project. See Giorgio Agamben, *The End of the Poem: Studies in Poetics*, trans. Daniel Heller-Roazen (Stanford, CA: Stanford University Press, 1999), 113. See his discussion of the *iustitium* provision in Roman law as an example of the light thrown by extreme cases (*SE*, 49).

6   Giorgio Agamben, *Remnants of Auschwitz: The Witness and the Archive*, trans. Daniel Heller-Roazen (1998; New York: Zone Books, 1999), 64–5.
7   He uses this view, for example, in *Homo Sacer* as a mode of argumentation. There he writes, "If today there is no longer any one clear figure of the sacred man, it is perhaps because we are all virtually *homines sacri*" (*HS*, 115).
8   Michel Foucault, *The History of Sexuality*, vol. 1, *An Introduction*, trans. Robert Hurley (New York: Vintage Books, 1978), 5–7.
9   See Foucault's *"Society Must Be Defended": Lectures at the Collège de France, 1975–1976*, ed. Mauro Bertani, trans. David Macey (New York: Picador, 2003), especially 26–28.
10 Michel Foucault, "On the Archaeology of the Sciences: Response to the Epistemology Circle," in *Aesthetics, Method, and Epistemology: Essential Works of Foucault, 1954–1984*, trans. Robert Hurley et al. (New York: New Press, 1998), 2:297–333, 2:303.
11 Gilles Deleuze, *Michel Foucault*, trans. Séan Hand (Minneapolis: University of Minnesota Press, 1988).
12 In 1971, Foucault, along with Daniel Defert, founded an activist group for prisoners, Groupe d'Information sur les Prisons (Prison Information Group). This group was unusual for having no goal beyond the documentation of what was intolerable in the system of penal incarceration; it explicitly detached itself from the reformist program of lobbying for an "ideal prison." Although the group had disbanded by 1974, Foucault saw its importance in terms of the experiment of removing activism from the justification of higher values. Are the terms of Agamben's future politics—the inoperativity of the law, the conception of gesture, and so on—"higher values" in this sense? Do they orient a particular understanding of the stakes of political action that suggests what should prevail beyond current conditions? That said, it seems to me that there is an important distinction to be made between Foucault's identification of the gap between the concepts of political philosophy and the modes of political activism and Agamben's thesis that the categories of political thought have become semantically impotent and that it is the process of this semantic exhaustion that takes precedence over political activism. The latter view is closer to quietism because of its destinal view of politics.
13 See Agamben's discussion of human rights in *Homo Sacer*, 126–36. The reasons for Agamben's position on human rights discourse must be seen as a consequence of the generality of his position on biopolitics. For a critique of Agamben that focuses on the proximity he shares with Arendt's view of a "pure" political domain, see Jacques Rancière, "Who Is the Subject of the Rights of Man?" *SAQ* 103.2–3 (Spring/Summer 2004): 297–310, especially 301–2.
14 See, for instance, Nancy's discussion of "Nietzsche's Age," in *The Gravity of Thought*, trans. François Raffoul and Gregory Recco (Atlantic Highlands, NJ: Humanities Press, 1997), 44.
15 In many places Agamben acknowledges his use of Nancy's understanding of the relation of the ban. However, there is also a shared diagnosis of our epoch as one of "accomplished nihilism," although the consequences each draws from this diagnosis are different. See Jean-Luc Nancy, *Birth to Presence*, trans. Brian Holmes (Stanford, CA: Stanford University Press, 1993), 44, and Agamben's use of the ban and discussion of nihilism in *Homo Sacer*, 29, 53.
16 Agamben, *The Open*, 87.

17 Agamben uses a discussion of Walter Benjamin to support his position in each of the cited passages.
18 I think one would need to look at Agamben's discussion of language in *The End of the Poem* and *Language and Death: The Place of Negativity*, trans. Karen E. Pinkus with Michael Hardt (Minneapolis: University of Minnesota Press, 1991), as well as his criticisms of philosophical aesthetics in *The Man without Content*, trans. Georgia Albert (Stanford, CA: Stanford University Press, 1999), in order to sketch out the position that supports this use of the arts.

Catherine Mills

Playing with Law: Agamben and Derrida on Postjuridical Justice

> Everything which is old, independent of its sacred origins, is liable to become a toy.
> —Giorgio Agamben, *Infancy and History*

> Once mislaid, broken and repaired, even the most princely doll becomes a capable proletarian comrade in the children's play commune.
> —Walter Benjamin, "Old Toys"

Toward the end of his book *State of Exception*, Giorgio Agamben writes:

> One day humanity will play with law just as children play with disused objects, not in order to restore them to their canonical use but to free them from it for good. What is found after the law is not a more proper and original use value that precedes the law but a new use that is born only after it. And use, which has been contaminated by law, must also be freed from its own value. This liberation is the task of study, or of play. And this studious play is the passage that allows us to arrive at that justice that one of Benjamin's posthumous fragments defines as a state of the world in which the world appears as a good that absolutely cannot be appropriated or made juridical.[1]

This comment follows a discussion of and reiterates a claim made by Walter Benjamin in his renowned essay on Franz Kafka: "The gate to justice is study."[2] Moving from Benjamin's formulation to his own conception of the future relation of humanity to law, Agamben appears to be offering a formulation of the means for disabling or rendering inoperative the "biopolitical machine" of modern political existence, even though that formulation is offered in the form of a futuristic prognosis rather than a prescription as such (SE, 64, 87–88). This affirmation concludes an extended discussion of the debate between Carl Schmitt and Benjamin, which Agamben addressed on prior occasions, including in *Homo Sacer*, the companion book to *State of Exception*. But Agamben's later discussion is interesting for its registration of distance from Schmitt and, correlatively, its more explicit indication of the Benjaminian loyalties that underpin Agamben's approach to questions of law, politics, and ethics. This is particularly the case with the appropriation of the idea of "studious play." But the recourse to the notion of playing with law as the path to justice introduces a number of questions that go unanswered within the text. What, for instance, does playing with law entail, and in what way might it lead humanity to a condition of justice after the law? What is the nature of this postjuridical justice, and what is its relation to the "happy life" (*ME*, 2–12, 114–15) that Agamben has previously identified as the necessary point of departure for political thought and existence?

This essay aims to provide answers to some of the questions raised by the formulation that Agamben introduces into his lexicon in *State of Exception*, of "playing" with law as the means for arriving at a postjuridical justice in which the world can no longer be appropriated by law. To do this, I return to several of Agamben's earlier texts, most notably, *Infancy and History* (*Infanzia e Storia*), which was first published in Italian in 1978 and translated into English in 1993.[3] It is here that Agamben most explicitly develops a conception of play, which he ties to the necessity of a reconsideration of history and the experience of prediscursive being that he calls *infancy*. My purpose in returning to this text is first to explicate the conceptual nexus of play, history, happiness, and justice and then to provide an account of the significance of playing with law in Agamben's work to date. In the second part of the essay, I build on this to briefly consider the theoretical and political implications of the conceptual apparatus at work here. It is worth noting from the outset that while I elaborate the conceptual crossovers between a number of Agamben's books, stretching from some of his earliest to the more recent, I do not mean to imply that there is a strong systematic coherence between all of his texts. In fact, there are a number of conceptual refractions and shifts that would be worth considering in a thorough exploration of this work. Such an undertaking is not possible here, though, and instead I offer a more general conceptual analysis of the later work in relation to the earlier.

On Play: Experience and History

State of Exception clarifies and extends the ontopolitical characterization of sovereignty, law, and legal violence that Agamben initially elaborates in *Homo Sacer*. In the earlier text, he describes the modern condition of law as one of "being in force without significance," a condition that is effectively equivalent to abandonment, wherein the subject of law is wholly given over to the violence of law and simultaneously bereft of its protection.[4] This condition emerges from the fact that the "state of exception" that founds sovereignty has become the rule, such that the law is suspended and yet remains in force. The later text elaborates the thesis, again beginning with the claim of an essential contiguity between the state of exception and sovereignty posited by Schmitt in *Political Theology*.[5] For Agamben, the general employment by law of the state of exception that this entails is the condition by which law simultaneously appropriates and abandons life. Within this, the unfounded decision of the sovereign is the "threshold," since it is in this decision that the originary noncoincidence between life and law is breached and life is truly brought into the sphere of law. However, in the chapter titled "Gigantomachy Concerning a Void," in which Agamben re-treats the Schmitt-Benjamin engagement, he suggests that the state of exception crucial to Schmitt is nothing other than a response to Benjamin's affirmation in "Critique of Violence" of a wholly anomic sphere of human action captured in the notion of a divine or pure violence.[6] Schmitt attempts, Agamben argues, to harness this sphere again to the operation of law, making it the topological condition of law's operation in suspension. Against this, Agamben affirms the importance of the distinction between the virtual and the real state of exception posited by Benjamin in his eighth thesis on the philosophy of history, which provides the means by which Benjamin can disable or thwart Schmitt's recuperative move. For the real state of exception is the anomic sphere of pure violence, of human action released from law and the instrumental violence it entails (*SE*, 52–64).

In general terms at least, this affirmation of the necessity of holding off the Schmittian recuperation of the anomic within law is consistent with the exhortation in *Homo Sacer* that the recognition of the status of law as being in force without significance in the ban is insufficient as the aim and achievement of contemporary thought, since residing in this recognition does little other than repeat the ontological structure of the sovereign ban. Against this, Agamben claims that contemporary thought must think abandonment beyond any conception of the law in order to move toward a politics freed of every ban. Thus, he writes in *Homo Sacer*, "Only where the experience of abandonment is freed from every idea of the law and destiny . . . is abandonment truly experienced as such" (*HS*, 60). While taking up the term *abandonment* in reference to Jean-Luc Nancy's essay "Abandoned Being," Agamben is thus quick to distance himself from Nancy's approach. While Nancy suggests that abandoned being must and can only respect the law, since "abandonment *is* abandonment to respect for the law,"[7] Agamben insists that abandonment must be pushed to its extreme limit in order to disable the law that is structured as a ban. In a move redolent of the Hölderlinian maxim that wherein lies danger also grows hope, for Agamben, it is at the extreme limit of abandonment that humanity is redeemed and the caesura of *bios* and *zoē* rendered obsolete; it is here that "happy life" finds its realization. However, it is at this juncture that the notion of playing with law takes on significance, since Agamben's proclamation of the need to push abandonment to its extreme limit raises the question — largely unanswered in the earlier discussion — of what can aid in the realization of this task.

As noted, the quote from *State of Exception* appears as the concluding paragraph of a discussion of the debate between Benjamin and Schmitt, a debate that Agamben has also addressed (though less directly) on prior occasions in *Homo Sacer* and *Potentialities*. However, the notion of play does not explicitly appear within the earlier discussions. Instead, what is at issue for Agamben is the "small adjustment" that allows the event of the Messiah to coincide with historical time but not be identifiable with it.[8] This adjustment is illustrated in Kafka's parable "Before the Law," in which Agamben sees the figuration of the messianic fulfillment of law in the closing of the door that completes the story (*HS*, 49–57). That is, the closing of the open door of the law allows for the restoration of meaning to the law and, therefore, its fulfillment, which amounts to the overturning of the Nothing that accompanies a law "in force without significance."[9] In the later discussion, though, the notion of the overturning of a law in force without significance gives way to the possibilities of humanity playing with law. What, then, is at stake in this shift or rephrasing? What is Agamben getting from the notion of *play* that was impeded in the earlier formulations of the messianic task? To gain an appreciation of this, it is first necessary to develop an understanding of the conception of play that is being invoked. To do this, I want to return to Agamben's early book *Infancy and History* and particularly to the chapter "In Playland," in which he analyzes the function of rituals and play in relation to time and claims that the revelatory characteristic of toys is to make present and tangible human temporality in itself.[10]

Agamben begins "In Playland" by citing Carlo Collodi's description of Playland in *Pinocchio*, in which a population entirely composed of boys partakes in all manner of games, creating a noisy and unconstrained pandemonium of play, the effect of which is to change and accelerate time and halt the repetition and alteration of the calendar. In contrast to play, Claude Lévi-Strauss provides an account of ritual in which the purpose or at least the effect of ritual is to "preserve the lived experience" and fix the alterations and repetitions of the calendar.[11] Between and from these positions, Agamben posits a relation of "correspondence and opposition" between ritual and play in regard to time and the calendar, wherein ritual "fixes and *structures*" the calendar, while play *"changes and destroys"* it (*IH*, 69). Indeed, play also has the capacity to undermine the sacred itself even while deriving from it. That is, while play is verified to derive from sacred practices, ceremonies, and rituals, it also transforms those practices to the point of destruction. Drawing on Émile Benveniste's study of play and the sacred, Agamben posits: "Playland is a country whose inhabitants are busy celebrating rituals, and manipulating objects and sacred words, whose sense and purpose they have, however, forgotten. . . . In play, man frees himself from sacred time and 'forgets' it in human time" (*IH*, 70). Even so, he continues, the relation of play and time is more specific, since the realm of the sacred does not exhaust that of play. Play also preserves profane objects and behaviors that have otherwise ceased to "exist," evident in the use that children make of objects that have outlasted their functional use value and are taken up again as toys. This "appropriation and transformation" is also apparent in the process of the "miniaturization" of objects that still maintain use value and circulate as functional tools—cars, electric cookers, airplanes, and so on (*IH*, 70–71). The dual mode of an object becoming a toy reveals the essential character of the toy as the "once upon a time" or "no more," where this temporal dimension is understood not only in a diachronic sense but also, as miniaturization shows, in a synchronic sense. Thus, "the toy is what belonged—*once, no longer*—to the realm of the sacred or of the practical-economic. . . . the essence of the toy is, then, an eminently historical thing; indeed, it is, so to speak, the Historical in its pure state" (*IH*, 71). The toy preserves of its sacred or economic model "the human temporality that was contained therein: its pure historical essence. . . . The toy is a materialization of the historicity contained in objects. . . . [It] makes present and renders tangible human temporality in itself, the pure differential margin between the 'once' and the 'no longer'" (*IH*, 72).

On the basis of this conception of the toy, Agamben suggests a further refinement in the apparently inverse functions of rite and play in relation to time: while rite annuls the disjuncture between a mythic past and the present and thereby reabsorbs diachrony into synchrony, play breaks any connection between past and present and transforms synchrony into diachrony. But, he suggests, it would be wrong to simply emphasize the distinction and disjuncture between rite and play, since empirically they are inseparable: "We always find play alongside ritual and ritual alongside play," and moreover, "every game . . . contains a ritual element and every rite an aspect of play" (*IH*, 74). This means, then, that rite and play should not be considered as having two distinct functions in relation to time; rather, both operate as "a single binary system, which is articulated across two categories which cannot be isolated and across whose correlation and difference the very functioning of the system is based" (*IH*, 74). The function of this binary system is to produce "a differential margin between diachrony and synchrony," and this differential margin is the condition of history, "in other words, human time" (*IH*, 75). That is to say, either the operation of play in the absence of the oppositional force of ritual or vice versa would result in the elimination of history in pure synchrony or diachrony. It is, then, the interaction of the poles of the "machine" of play and ritual that produces history as the irreducible residue of the impossibility of a pure transformation of synchrony into diachrony or diachrony into synchrony.

At this point, it is worth noting that while Agamben is ostensibly interested in the notion of play in "In Playland," his focus quickly shifts from the interactive activity of playing to the ontological status of toys, themselves understood simply as objects of everyday life that have become elements of play. This narrowing of focus is undertaken without justification and has the effect of eliding analysis of toys that have no immediate reference point in everyday life as either disused objects or objects of miniaturization, as well as those ways of playing that do not revolve around toy objects at all but are instead structured by more or less arbitrary rules of the game. It is also nostalgic, such that it is difficult to see what bearing this conception may have on contemporary modes of play, which often involve sophisticated technologies and "virtual realities" that may or may not contain human beings in their character sets.[12] In spite of the descriptive limitations of this conception of play, it nevertheless fulfills an important role in the conceptual nexus that Agamben develops in *Infancy and History*. In particular and most obviously, it is integral to his reconsideration of temporality and history, understood more specifically as the historicity of the human. Less obviously, but no less important, it is tied to a reflection on experience and particularly on the "originary" prediscursive experience of infancy.

This conceptual nexus is made explicit in the eponymous chapter of *Infancy and History*, which takes as its central theme the destruction of experience in the modern world. Agamben posits the poverty and denial of experience in everyday life in the modern world, but also maintains that this is not a cause for despair. Rather, the apparent denial of experience may well provide the ground for a "germinating seed" of a future experience. In attaining toward a new conception of experience, he ultimately suggests this can be found in "infancy," understood as a wordless, mute condition that precedes speech, where this precedence is not chronological but ontological, such that infancy coexists with language and is expropriated by it in the constitution of the subject. Infancy is the experience from which the human subject emerges, that being for whom language is not an essential possession but a deprivation insofar as an (expropriating) appropriation of language is required in order to speak. That is, "man" must constitute himself as speaking subject, specifically through the appropriation of the personal pronoun and, in doing so, falls away from the originary, "transcendental" experience of infancy. In this, infancy makes evident a point of transition in the constitution of the human as speaking subject, a transition that Agamben argues is best understood in terms of the distinction between the semiotic and semantic introduced by Benveniste. The nonintersection of these two spheres requires a transformation of the "pure pre-babble language of nature" of the semiotic into discourse, a transformative moment that, for Agamben, is intrinsically related to the human constitution of itself as subject of language. He concludes, "The human is nothing other than this very passage from pure language to discourse; and this transition, this instant, is history" (*IH*, 55). Thus, infancy is posited as the origin of history, since the condition of possibility of history is the discontinuity between language and discourse that infancy introduces and of which it is the essential experience. Further, this indicates that the human is by nature historical, since the human is itself nothing apart from the transition from language to speech that conditions history.

The implication of reading "In Playland" and "Infancy and History" together is that the revelation of the historicity of the human that infancy constitutes (in setting up a break between language and discourse) occurs through the intersection of rite and play (insofar as their interaction produces a differential margin between diachrony and synchrony).[13] If this is accurate, more needs to be said of the conception of history that Agamben is developing, since it cannot be—and is not—based on the traditional notion of history as the (record of) the linear succession of instants. Without delving far into Agamben's philosophy of history, it should come as no surprise that it is at least partly motivated by Benjamin's exhortation in his theses on history to attain toward an understanding appropriate to the real state of exception, that is, to the "messianic kingdom."[14] Thus, Agamben draws on the concept of the *cairos* and develops an understanding akin to Benjamin's notion of the "now-time," or *Jetztzeit*, as well as Martin Heidegger's concept of *Ereignis*, in which history is an originary characteristic of man and is revealed in an irruption that breaks apart inauthentic chronological time. As Agamben writes, "History . . . cannot be the continuous progression of speaking humanity through linear time, but in its essence is hiatus, discontinuity, epoché" (*IH*, 53). Further, in a view that recalls the status of happiness in Benjamin's "Theologico-Political Fragment,"[15] Agamben argues that such a conception of history is intimately related to "humankind's original home" of pleasure, so much so that pleasure is "the true site of history," and the "primary core of all authentic historical experience" is Adam's seven paradisiacal hours. Within this, "history is not . . . man's servitude to continuous linear time, but man's liberation from it: the time of history and the *cairos* in which man, by his initiative, grasps favourable opportunity and chooses his own freedom in the moment" (*IH*, 104). Thus, "the chronological time of pseudo-history must be opposed by the *cairological* time of authentic history." Agamben concludes from this that the "true revolutionary and the true seer" is "he who, in the epoch of pleasure, has remembered history as he would his original home" because he is "released from time not at the millennium, but now" (*IH*, 105).

To return to my starting point, more can now be said of the idea of playing with law as if it were a disused object, that is, a toy. It is now possible to better appreciate the perceived revolutionary potential of play and of the toy. As we have seen, the toy brings to light the "temporality of history in its pure differential and qualitative value." That is, in making present "human temporality in itself, the pure differential margin between the 'once' and the 'no longer'" (*IH*, 72), the toy permits a release from continuous and linear time and the realization of and return to history, understood as the true homeland of humanity (*IH*, 104–5). In relation to law, we can now say that as a disused object the law has lost its use value in the realm of the politico-economic and has instead been relegated to the profane use that can be made of it by children. The characterization of its being in force without significance appears to locate the law within the diachronic element of the "'once' . . . 'no longer,'" rather than within the synchrony of miniaturization. This is significant because it highlights the ritualistic dimension of law, which compensates for the disjuncture of past and present, Agamben argues, by reabsorbing diachrony into synchrony. Play, however, transforms synchrony into diachrony by breaking the tie between past and present. This production of a differential margin in the dialectic of rite and play is the condition of history; it is that which allows for the now. As a toy and only as a toy, as an object of play, the rite of law contributes to the revelation of the essential historicity of the human.

The ritualistic dimension of law is important for another reason as well. Agamben insists on the impossibility of the elimination of either diachronic or synchronic signification: in all games and rites, the one remains a stumbling block for the other, thereby preventing the attainment of a pure state of diachrony or synchrony. Thus, he writes, "at the end of the game," the toy—the privileged signifier of absolute diachrony—"turns around into its opposite and is presented as the synchronic residue that the game can no longer eliminate" (*IH*, 79). This implies that playing with law does not mean eliminating the law, for there is actually a sense in which the law is rescued from its own obsolescence in play. Rather than being maintained solely in a state of decay characterized by the simple lack of practico-economic value as law, it is given a new use. But this does not take the form of a resacralization of the law and restoration of transcendental meaning or force. Instead, the new use of law takes the form of its deactivation or deposition. Before saying more of this, it is worth cautioning against the phrase "at the end of the game" used above, for in what sense would the game in which humanity plays with law have an end? To construe the game of playing with law as having an end would in fact push Agamben's conception of the messianic toward an identification with the eschatological, a conflation that he explicitly resists in *The Time That Remains*.¹⁶ Thus, within his own characterization, it would be more accurate to insist on the endlessness of play. As with the activity of study with which it is intimately related in the paragraph in question, play is interminable; it has no end beyond pleasure. As Agamben writes in *Idea of Prose*, "Not only can study have no rightful end, it does not even desire one."¹⁷

In fact, it is presumably the endlessness of play that allows for the noninstrumental appropriation of law and ultimately its deactivation in play; that is, the "free use" of law within play exceeds the constraints of instrumentality and gives onto a justice that Agamben identifies as akin to a condition in which the world can no longer be appropriated by law. In this way, the noninstrumentality and interminability of play ensure a passage to a justice that is irreducible to law. As Agamben writes, "The law—no longer practiced but studied—is not justice, but only the gate that leads to it. What opens a passage toward justice is not the erasure of law, but its deactivation and inactivity—that is, another use of law" (*SE*, 64). One of the questions that this raises is to what extent a deposed or deactivated law remains a law. In what sense is a deposed law still a law? Agamben suggests that it is this question of the status and meaning of law after its messianic fulfillment that motivates Benjamin's reflections on Kafka, in which law "no longer has force or application" (*SE*, 63). However, this raises more questions than it answers, and in particular, it leaves open what a postjuridical justice arrived at through studious play might look like. We can be sure that what Agamben means by "justice" does not coincide with more standard jurisprudential conceptions as the proper application of law. Despite his concern with questions of law, though, the concept of justice has played a small part in Agamben's work to date (at least if considered at an explicit textual level), and there is little overt indication of what it would amount to beyond this discussion in *State of Exception*. One point at which a (slightly) more extended consideration of justice does appear is in an early fragment in which Agamben defines justice as "the handing on of the Forgotten" and the "transmission of oblivion" (*IP*, 79). At first glance, this does little to clarify the concept of justice that he employs, but it does point toward a path of elucidation.

Agamben's conception of justice is indebted to both an opposition between justice and the natural world and the distinction between justice and profane law found in Judaism, particularly as these are rendered through the interpretive matrix delineated by Benjamin and Kafka. The triad of law, justice, and the natural order is provocatively addressed by Benjamin in several crucial essays, not least "Critique of Violence," in which he posits the necessity of a divine violence that destroys mythic, legal violence and expiates the guilt of "mere life." Resisting the reduction of "man" to mere life that the modern conception of the sanctity of life threatens, Benjamin begins to isolate the "not-yet-attained" condition of the just man, who must necessarily exist outside the realm of fate that underlies all manifestations of legal violence. In this figuration, justice serves to distinguish the sacredness of man from the natural life expressed in plants and animals, since if there is anything sacred in man, it could not be the shared fact of biological existence. Extending this theme in "Franz Kafka," Benjamin argues that Kafka's characters are mired in a primeval guilt-laden "swamp world" ("FK," 808).[18] This world, he suggests, has been forgotten, but this does not in any way mean that it does not impact on the present; on the contrary, "it is present by virtue of this very oblivion," and oblivion is the receptacle from which this "inexhaustible, intermediate world" presses forth ("FK," 814). Benjamin goes on to characterize the force of forgetting as a "tempest" and privileges the activity of study as "a cavalry attack against it," thus moving to the conclusion that Agamben reiterates in *State of Exception*: "The gate to justice is study" ("FK," 815).

Significantly, Agamben does not attend to these references to the Forgotten in Benjamin's interpretation of Kafka in appropriating the idea of study in *State of Exception*. At the least, this indicates that the idea of the Forgotten that he suggests in the fragment on justice is not simply an invocation of the primeval world that Benjamin reads in Kafka, nor is it, for that matter, strictly analogous to the conception of justice entailed by Benjamin's "history of the oppressed," a phrase that Agamben does not take up in his reading of Benjamin's theses on history.[19] Rather, the "Forgotten" refers to the element in speech that is necessarily unspeakable, that which cannot pass into discourse but is instead located solely in the voice "like a heralding gesture or vocation" (*IP*, 79). Or, as Agamben puts it in *Remnants of Auschwitz*, it is the unspoken in any appropriation of language, that to which the speaking subject inescapably bears witness.[20] *Logos* and *dike* are, then, "indistinguishable from the start" (*IP*, 80). Of this understanding of justice, two features are especially significant. First, in its relation to the Forgotten of language, justice is distinct from law without being simply opposed to it: justice and law are not immiscible as are oil and water; rather, the former is insuperably part of the latter. Further, it is only because justice is found within law that law can contribute to the attainment of justice, though it is not justice itself.[21] The second important feature is that justice requires the *transmission* of the Forgotten—that is, only in its transmission does the Forgotten relate to justice. This clarifies the significance of "bearing witness" to the unspeakable that resides within speech in the elaboration of justice as ethos that Agamben develops in *Remnants*. It also clarifies the role of play further, for studious play now appears as the means of transmission of the Forgotten. And as the means of transmission, play purifies—and not simply of guilt but, as Benjamin suggests, of law—thereby returning humanity to the just world of the Forgotten or oblivion, not once and for all in the *eschaton*, but endlessly.

It is not hard to see that the condition of justice that humanity is returned to in playing with law is tied to the experience of infancy that Agamben proposes as the origin of human historicity, that which humanity must "continually travel toward and through" (*IH*, 53). In play, humanity is returned to a "being-in-language"[22] in which what is at stake is communicability itself, that is, potentiality. In his historical condition, man lives "happily," where the idea of happiness "takes up" the "what has never been" (*IP*, 93) of a life or the unlived potentiality that resides in every gesture and act, no matter how habitual or conventional. Happy life is the "beatitude"[23] that resides inexorably in every life as the unlived potentiality of its own composition. Resonant of Benjamin's suggestion that happiness is a condition of fatelessness (happiness, he suggests, "releases man from the embroilments of fate and from the net of his own fate"[24]), this happy life or "form-of-life" is a life that has "reached the perfection of its own power and of its own communicability—a life over which sovereignty and right no longer have any hold."[25] This "life of power" (*ME*, 9) gives onto a new communism, in which nothing is shared except the power and possibility of life itself, and life escapes the caesuras and impotence to which law has relegated it. As articulated in *The Coming Community*, then, Agamben's vision of political futurity turns around the nonidentitarian unification of life with its own potentiality or "being-thus."[26] Importantly, this new communism is not something to be invented or found in the future, for the coming community exists now as a community of "immediate linguistic and ontological transparency"[27] without identity and sharing nothing except the being-thus of happy life, in which all belong without any claim to belong. Moreover, it is precisely the nonidentitarian nature of the coming community that opposes it to the state and state political forms, since ultimately the state can recognize any claim to identity but "cannot tolerate . . . that singularities form a community without affirming an identity, that humans co-belong without any representable condition of belonging."[28]

Postjuridical Justice: "The Origin and End of Play"

At this point, I want to consider some of the implications of the foregoing account of law and justice, particularly with an eye to the practico-theoretical conclusions that can be drawn from within this framework. Undoubtedly, any such effort is beset by problems from the start, not least because the historiographical and philosophical method that Agamben relies on runs dangerously close to simultaneously crediting "a logic of the empirical event" while discrediting the same logic "in the name of [a] trans-historic and natural ideal," that is, of relying on a "sleight-of-hand trick between history and nature, between historical empiricity and teleological transcendentality."[29] While clearly not committed to a liberal democratic teleology in the manner of Francis Fukuyama—the target of Jacques Derrida's comment—the relation between the empirical and theoretical within Agamben's historiography can certainly be questioned. Further, there is some relevance in criticisms that echo—wittingly or not—Jürgen Habermas's comment on Benjamin that his "rescuing critique . . . has a highly mediated position relative to political praxis" since "liberation from cultural tradition of semantic potentials . . . is not the same as the liberation of political domination from structural violence."[30] Similarly, for many, Agamben's emphasis on the messianic fulfillment of law is a deeply unsatisfying political response to the structural violence of the current global order.[31] However, the difficulty that such criticisms rarely register is that Agamben puts into question the very notion of "use" and its value and, therefore, the interrelation of theory and praxis presupposed by the charge. That said, it remains the case that his work has currency as a diagnostic framework for approaching aspects of the current juridico-political order, and by the evidence of his own interventions, Agamben also apparently intends it to have such a bearing. It is on the basis of this that I will explore some of its practico-theoretical implications.

In doing so, it is first instructive to consider Agamben's conception of justice alongside that elaborated by Derrida, since both engage with the triad of conceptions of law and legal violence offered by Benjamin, Kafka, and Schmitt. My aim in approaching the long-standing "conversation" between Agamben and Derrida is not to provide a full exploration of either the "esoteric" or "exoteric" dossier—to adopt Agamben's terms momentarily—of the engagement.[32] Instead, I draw on Derrida heuristically, to isolate and illustrate several implications of Agamben's claims. Derrida's concern with questions of justice and legal violence stretches back at least to "Force of Law," where he distinguishes between law and incalculable justice through a reading of "Critique of Violence."[33] In his essay, Derrida explicitly links the calculability of the law with the impossible but necessary experience of the aporias that condition the political decision. In identifying a necessary residue of the undecidable in every political decision, Derrida argues that an infinite, irreducible "idea of justice" haunts every decision and necessarily haunts it in order for it to be a decision and not merely the application of a rule. In the face of this undecidability, though, Derrida also insists on the ongoing urgency of the decision, since incalculable justice requires calculation—it requires that the decision on what is just and right be made at any moment. As he writes, "Not only must we calculate, negotiate the relation between the calculable and the incalculable . . . but we must take it as far as possible, beyond the place we find ourselves and beyond the already definable zones of morality or politics or law" ("FL," 28).

Of the irreducible idea of justice, Derrida goes on to say that he would hesitate to "assimilate too quickly this 'idea of justice' to a regulative idea (in the Kantian sense), to a messianic promise or to other horizons *of the same type*" ("FL," 25).[34] Derrida explains this hesitancy further, suggesting that his reason for keeping distance from the Kantian regulative idea or messianic advent is precisely because they are *horizons*, a term that indicates both "the opening and the limit that defines an infinite progress or period of waiting." Thus, Derrida rejects the formulation of the messianic advent as waiting, as requiring infinite patience, because justice "does not wait." Incalculable justice does not require patience but a just decision "right away" as he says ("FL," 26). It necessitates the negotiation between calculable law and incalculable justice as an urgent decision, but, at the same time, justice is the irreducible "to-come" of the decision. As Derrida writes, justice "has no horizon of expectation . . . But for this very reason, it *may* have an *avenir*, a 'to-come'. . . . Justice remains, is yet, to come, *à venir*, it has an, it is *à-venir*, the very dimension of events irreducibly to come. . . . Justice as the experience of absolute alterity is unpresentable, but it is the chance of the event and the condition of history" ("FL," 27). Justice does not entail a horizon of waiting, since the "to-come" of justice "is not a horizon but the disruption or opening up of the horizon,"[35] which necessitates constant and urgent engagement with the aporias of undecidability. Derrida's insistence on the interminable urgency of the undecidable marks his distance from Benjamin, made more explicit in the postscript to "Force of Law." In criticizing Benjamin's claim of the necessity of an expiatory divine violence, Derrida suggests that such a conception of messianism is haunted by "the theme of radical destruction, extermination, total annihilation, beginning with the annihilation of the law and right, if not of justice" ("FL," 63, n. 6). Whether Derrida's reading of Benjamin is justifiable or not, the distance between Derrida and Agamben is clear: whereas for the former a radically destructive divine violence risks eliminating the possibility of justice along with law and right, for the latter divine violence is the necessary condition of justice.

But Derrida's insistence on undecidability is important here for reasons other than this comparative point. One of the most striking characteristics of Derrida's conception of justice is its engagement with Emmanuel Levinas's insistence on the primordiality of alterity and correlative positioning of ethics as first philosophy. I cannot address the complexity of this philosophical debt here, but suffice it to note the association of justice with alterity, in the claim that it is *"as the experience of absolute alterity"* ("FL," 27, my emphasis) that justice is the condition of history and chance of the event. It is as the experience of alterity that incalculable justice renders the decision impossible though urgent and the "to-come" of justice heralds the impossible presence of the Other, the absolute other, or *arrivant*.[36] But if the concern with justice emerged in Derrida's work in 1989, alterity and difference were long-standing dimensions of deconstruction rendered through the "concepts" of *différance* and the trace.[37] Both of these work to disrupt—though not overcome—the history of the metaphysics of presence and corollaries such as origin, causality, and end (in the sense of both *telos* and *eschaton*) on which such a metaphysic is premised. Understood as the "play of differences," *différance* constantly and necessarily defers the originary, not least because play is itself nothing other than deferral and indetermination, announcing as it does "on the eve of philosophy and beyond it, the unity of chance and necessity in calculations without end."[38] Significantly, Derrida is explicit that *différance* does not just pertain to the distinction within language between the signifier and signified; rather, he suggests *différance* is "*also the relation of speech to language,* the detour through which I must pass in order to speak."[39] Thus, for Derrida, the passage from speech to language is itself riven by the play of difference, by deferral and indetermination without end. While this portrayal barely touches on the nonoriginary role of difference within Derrida's work and its influence on his conception of a justice that is irreducible to calculable law and politics, it is sufficient to make my point in relation to Agamben.

The notion of playing with law offers Agamben a path between, on the one hand, the endless deferral of deconstruction that "in maintaining law in a spectral life, can no longer get to the bottom of it" (*SE*, 64) and, on the other, eschatological or prophetic conceptions of the end of time. Similarly, the coming community of whatever being is posed as a way of overcoming or surpassing the philosophical and political dilemmas of identity and difference. But do these philosophical moves achieve what they set out to achieve, or do they fall short of their measure and reinstitute the very problems that they aim to address at another level? The question of whether Agamben's formulation of justice beyond law rests on the reinstitution of ontological unity and full presence is not one I can address in full detail here. But consider that the vision of a postjuridical justice predicated on a return to a kind of prelapsarian experience of infancy entails an absence of alterity at the origin of the human. That is, the experience of infancy amounts to a return to a univocality that ontologically precedes the differentiating and identificatory power of language as discourse or systems of signification. Further, analogous to infancy, the condition of happy life is life lived beyond differentiation and distinction, and particularly the fracturing of life into component parts—biological, political, intellectual, and so on—such that an integrated happy life, or form-of-life, is said to provide the "unitary centre" of the coming politics (*ME*, 12). To be sure, the conception of origin that Agamben is working with is not straightforward—he states that since the originary experience of infancy is born in and institutes the caesura of language and discourse, "man's nature is split at its source" and is, therefore, historical (*IH*, 49). But while he distances himself from a conception of origin as foundational in a straightforward sense and from the identification of the messianic with the eschatological, this thought is not free of invocations of the tropes of "end and origin," which serve to undercut the clarity of his excision. This is much more the case in his earlier work, in which he resists a romanticism of the prelapsarian and *eschaton* with less rigor, but such invocations are not wholly absent from the later political philosophy either.

One further consequence of the elimination of claims to identity that Agamben's coming politics and community entail is that it also requires overcoming relationality—and Agamben states that it "implies nothing less than thinking ontology and politics beyond every figure of relation" (*HS*, 47). As I have argued elsewhere, to the extent that relationality enters into Agamben's thought, it does so only in the form of autoaffection.[40] But this radical rejection of relationality appears to be premised on the elimination of alterity altogether, including the alterity internal to and constitutive of autoaffection. The question that can be asked here, then, is what the cost of this would be. To make clear the weight of this question, let me briefly consider one way in which this neglect of alterity and difference plays out. As I mentioned previously, Agamben takes his initial inspiration for his conception of play as an interruption of calendrical time from Collodi's description of Playland in *Pinocchio*. One notable characteristic of this description that Agamben passes over in silence is that Playland is populated entirely by boys. Without taking this issue up in detail, the question that poses itself here is to what extent this idyllic conception of play is actively premised on the exclusion of gender difference. In what way would the presence of a girl or girls disrupt, interrupt, or undo the cacophonous, indifferent play of boys? Not dissimilarly, Agamben is silent on issues of gender in his reference to Aristotle's distinction between the life of the *oikos* and politics, even though it is insistently present in the designation of the *oikos* as the domain of reproduction that necessarily precedes and supports the life of politics. As Derrida sharply remarks, the distinction of *bios* and *zoē* is not as straightforward as Agamben takes it to be.[41] The point here is not that "girls" or "women" should, as if they could, simply be added to the scene of play or biopolitics in such a way that the scene itself and the conceptual framework built on it would remain without substantial change. Nor is the point to simply note the exclusion of women from Agamben's philosophical lexicon at an explicit textual level—the consistent use of gender-specific pronouns as if their reference were universal may well be indicative of a philosophical blindness or "amnesia,"[42] but it does not reach to the depths of the problem in itself. For what would it be to ask the "question of gender" within the messianic framework that Agamben proposes? Indeed, can such questions be asked within that framework?

Curiously, one other point at which Agamben veers toward—though without in any way taking up—questions of gender is in a fragment titled "The Idea of Communism."[43] The primary focus of this fragment is pornography, in which the utopia of a classless society is said to appear in the "gross caricatures" (*IP*, 73) of all markers of class and their transformation in the sexual act, itself the necessary conclusion of any pornographic film. The "eternal political justification," he argues, for pornography is its capacity to reveal the presence of pleasure in everyday life, even if the pleasure that it brings to light is only temporary and fleeting. Pornography "does not elevate the everyday world to the everlasting heaven of pleasure" and necessarily remains limited to revealing the "inner aimlessness of every universal." He goes on to conclude that "pornography achieves its intention" in "representing the pleasure of the woman, inscribed solely in her face" (*IP*, 74). What is one to make of this figuration of woman within the domain of the ephemeral pleasure of everyday life, unable to move from that to the "everlasting heaven of pleasure"? There is of course a long philosophical tradition of casting women as the privileged figure of ephemerality, unable to gain or yield access to the universal. What is more specific to Agamben, though, is that this incapacity to attain toward the universal implies that women cannot enter in the "homeland of humanity," the "everlasting pleasure" of happy life, and the condition of historicity. More specifically, women cannot enter in the homeland of humanity as women, since that entrance requires the abolition of all characteristics of identity. It is not as women that women can enter into everlasting happiness but only as "whatever." But this relegation of all identifying characteristics to the traps of chronological time, discourse (as opposed to vocality and the experience of infancy), and control by the state is surely problematic, even when one does not favor an identitarian politics instead. For one, the romanticized ideal of a radically immanent, unified life beyond all identity precludes analysis of the various regimes of identification and disidentification that currently operate to establish the different political and ethical valuations of lives in their manifold expression. This moves Agamben's conceptions of politics and justice too far from an appreciation of the unequal imposition of burdens and vulnerabilities within a globalized biopolitical order. Difficult and dangerous as the political negotiation of alterity surely is in this context, the best response is not the obviation of difference and identity.

To conclude, then, it is perhaps not without significance that one of the primary resources that Agamben draws on in his discussion of play is Lévi-

Strauss. "In Playland" pays homage to the anthropologist and adopts his formulation of the effects on calendrical time of ritual and play, while correcting the study of the *churinga*, developed in *The Savage Mind*, and is dedicated to Lévi-Strauss for the occasion of his seventieth birthday. The chapter is chronologically preceded by Derrida's own engagements with Lévi-Strauss, particularly the analysis that is reiterated in *Of Grammatology* and *Writing and Difference*. In "Structure, Sign, and Play," Derrida offers important insight into the concept of play when he identifies two tensions that the concept of play cannot be separated from in Lévi-Strauss's work, the first of which is in its relation to history and the second in relation to presence. On the basis of these, Derrida concludes that the overriding conception at work in Lévi-Strauss is a "saddened, negative, nostalgic, guilty, Rousseauistic" approach to play that is "turned toward the lost or impossible presence of the absent origin," a play that "escapes the order of the sign" and is motivated by the reinstitution of the center as foundation and assurance of presence.[44] Would it be wholly inaccurate to suggest that this conception of play is also found in Agamben's work? Certainly, his conception of play is not turned toward a lost past as the condition that chronologically precedes the degradations of culture and language. But what is at stake in play is the essential historicity of the human, the "homeland of humanity," revealed through the originary experience of infancy, an experience that ontologically precedes and institutes the caesura of language and signification. In the face of this, it may be worth affirming Derrida's description of a second—though not strictly opposed—"Nietzschean" conception of play, of which he writes that this way of "thinking of play" entails a "joyous affirmation of the play of the world and of the innocence of becoming, the affirmation of a world of signs without fault, without truth, and without origin.... This affirmation then determines the noncenter otherwise than as a loss of center. And it plays without security," no longer dreaming of "full presence, the reassuring foundation, the origin and end of play."[45]

Notes

1 Giorgio Agamben, *State of Exception*, trans. Kevin Attell (Chicago: University of Chicago Press, 2005), 64. Hereafter cited parenthetically by page number as *SE*.
2 Walter Benjamin, "Franz Kafka," trans. Harry Zohn, in *Walter Benjamin: Selected Writings*, vol. 2, *1927–1934*, ed. Michael W. Jennings, Howard Eiland, and Gary Smith (Cambridge, MA: Belknap Press, 1999), 815. Hereafter cited parenthetically by page number as "FK."

3   Giorgio Agamben, *Infancy and History: Essays on the Destruction of Experience*, trans. Liz Heron (London: Verso, 1993). Hereafter cited parenthetically by page number as *IH*.
4   Giorgio Agamben, *Homo Sacer: Sovereign Power and Bare Life*, trans. Daniel Heller-Roazen (Stanford, CA: Stanford University Press, 1998), 51 passim. Hereafter cited parenthetically by page number as *HS*.
5   Carl Schmitt, *Political Theology: Four Chapters on the Concept of Sovereignty*, trans. George Schwab (Cambridge, MA: MIT Press, 1985), esp. 12–15.
6   Giorgio Agamben, "Gigantomachy Concerning a Void," *SE*, 52–64; and Walter Benjamin, "Critique of Violence," trans. Edmund Jephcott, in *Walter Benjamin: Selected Writings*, vol. 1, *1913–1926*, ed. Marcus Bullock and Michael W. Jennings (Cambridge, MA: Belknap Press, 2003), 236–52.
7   Jean-Luc Nancy, "Abandoned Being," in *The Birth to Presence*, trans. B. Holmes et al. (Stanford, CA: Stanford University Press, 1993), 36–47, at 44.
8   Giorgio Agamben, "The Messiah and the Sovereign: The Problem of Law in Walter Benjamin," in *Potentialities: Collected Essays in Philosophy*, ed. and trans. Daniel Heller-Roazen (Stanford, CA: Stanford University Press, 1999), 160–74, at 172–74.
9   For a further discussion of Agamben's reading of this parable, see Catherine Mills, "Agamben's Messianic Politics," *Contretemps* 5 (December 2004), www.usyd.edu.au/contretemps/5december2004/mills.pdf (accessed July 11, 2007).
10 Giorgio Agamben, "In Playland: Reflections on History and Play," in *IH*, 65–87.
11 Claude Lévi-Strauss, *The Savage Mind* (London: Weidenfeld and Nicolson, 1972), esp. 17–32.
12 But see Adrian Mackenzie, *Transductions: Bodies and Machines at Speed* (New York: Continuum, 2003).
13 Giorgio Agamben, "Infancy and History: An Essay on the Destruction of History," in *IH*, 11–63.
14 See Walter Benjamin, "Theologico-Political Fragment," trans. Edmund Jephcott, in *Walter Benjamin: Selected Writings*, vol. 3, *1935–1938*, ed. Howard Eiland and Michael W. Jennings (Cambridge, MA: Belknap Press, 2002); and "On the Concept of History," trans. Harry Zohn, in *Walter Benjamin: Selected Writings*, vol. 4, *1938–1940*, ed. Howard Eiland and Michael W. Jennings (Cambridge, MA: Belknap Press, 2003).
15 Benjamin, "Theologico-Political Fragment," 305–6. Also see Benjamin's comment in "On the Concept of History": "The idea of happiness is indissolubly bound up with redemption. The same applies to the idea of the past, which is the concern of history" (389).
16 Giorgio Agamben, *The Time That Remains: A Commentary on the Letter to the Romans*, trans. Patricia Dailey (Stanford, CA: Stanford University Press, 2005), 62.
17 Giorgio Agamben, *Idea of Prose*, trans. Michael Sullivan and Sam Whitsitt (Albany: State University of New York Press, 1995), 64. Hereafter cited parenthetically by page number as *IP*.
18 For a further discussion of this essay, see Beatrice Hanssen, *Benjamin's Other History: Of Stones, Animals, Human Beings, and Angels* (Berkeley: University of California Press, 1998), esp. 137–49.
19 For a detailed discussion of this dimension of Benjamin's thought, see Matthias Fritsch, *The Promise of Memory: History and Politics in Marx, Benjamin, and Derrida* (Albany: State University of New York Press, 2005).

20 Giorgio Agamben, *Remnants of Auschwitz: The Witness and the Archive*, trans. Daniel Heller-Roazen (New York: Zone Books, 1999).
21 See G. W. F. Hegel, *Phenomenology of Spirit*, trans. A. V. Miller (Oxford: Oxford University Press, 1977), esp. 23, for further discussion of the kind of interrelation that Agamben appears to be proposing.
22 Giorgio Agamben, *The Coming Community*, trans. Michael Hardt (Minneapolis: University of Minnesota Press, 1993), 87 passim.
23 See Giorgio Agamben, "Absolute Immanence," in *Potentialities*, 220–39; see also *IP*, 93.
24 Walter Benjamin, "Fate and Character," trans. Edmund Jephcott, in *Selected Writings*, vol. 1, 203.
25 Giorgio Agamben, *Means without End: Notes on Politics*, trans. Vincenzo Binetti and Cesare Casarino (Minneapolis: University of Minnesota Press, 2000), 114–15. Hereafter cited parenthetically by page number as *ME*.
26 Agamben, *The Coming Community*, 93 passim.
27 Adam Thurschwell, *Specters of Nietzsche: Potential Futures for the Concept of the Political in Agamben and Derrida*, www.law.csuohio.edu/faculty/thurschwell/nietzsche.pdf, 26 (accessed July 11, 2007).
28 Agamben, *The Coming Community*, 86.
29 Jacques Derrida, *Specters of Marx: The State of the Debt, the Work of Mourning, and the New International*, trans. Peggy Kamuf (New York: Routledge, 1994), 69. Of course, the term *history* is used here with a different sense than that given to it by Agamben.
30 Jürgen Habermas, "Walter Benjamin: Consciousness-Raising or Rescuing Critique," in *On Walter Benjamin: Critical Essays and Recollections*, ed. Gary Smith (Cambridge, MA: MIT Press, 1988), 90–128, at 118, 120.
31 See, for instance, Robert Sinnerbrink, "From *Machenschaft* to Biopolitics: A Genealogical Critique of Biopower," *Critical Horizons* 6.1 (2005): 239–65.
32 See Thurschwell, *Specters of Nietzsche*, for a more extended comparison of Agamben and Derrida.
33 Jacques Derrida, "Force of Law: The Mystical Foundations of Authority," in *Deconstruction and the Possibility of Justice*, ed. Drucilla Cornell, Michel Rosenfeld, and David Gray Carlson (New York: Routledge, 1992), 3–67. Hereafter cited parenthetically by page number as "FL."
34 For further discussion of this comment and its later reformulation, see John D. Caputo, *The Prayers and Tears of Jacques Derrida: Religion without Religion* (Bloomington: Indiana University Press, 1997), 117.
35 Ibid., 118.
36 Also see Jacques Derrida, *Of Hospitality* (Stanford, CA: Stanford University Press, 2000).
37 See Robert Bernasconi, "The Trace of Levinas in Derrida," *Derrida and Différance*, ed. David Wood and Robert Bernasconi (Evanston, IL: Northwestern University Press, 1988), 13–29, for a discussion of Derrida's understanding of the trace in relation to Levinas.
38 Jacques Derrida, *Margins of Philosophy*, trans. Alan Bass (Chicago: University of Chicago Press, 1982), 7.
39 Ibid., 15, my emphasis.
40 See Catherine Mills, "Linguistic Survival and Ethicality: Biopolitics, Subjectification, and

Testimony," in *Politics, Metaphysics, and Death: Essays on Giorgio Agamben's "Homo Sacer"*, ed. Andrew Norris (Durham, NC: Duke University Press, 2005), 198–221.
41 Jacques Derrida, *Rogues: Two Essays on Reason*, trans. Pascale-Anne Brault and Michael Naas (Stanford, CA: Stanford University Press, 2005), 24.
42 Adriana Cavarero, "Equality and Sexual Difference: Amnesia in Political Thought," in *Beyond Equality and Difference: Citizenship, Feminist Politics, and Female Subjectivity*, ed. Gisela Bock and Susan James (New York: Routledge, 1992), 32–47.
43 Giorgio Agamben, "The Idea of Communism," in *IP*, 73–75.
44 Jacques Derrida, "Structure, Sign, and Play," *Writing and Difference*, trans. Alan Bass (Chicago: University of Chicago Press, 1978), 278–293, at 292.
45 Ibid.

Eleanor Kaufman

The Saturday of Messianic Time
(Agamben and Badiou on the Apostle Paul)

There has been a striking theological turn in contemporary continental thought over the past decade. Philosophers ranging from Jacques Derrida to Jean-François Lyotard to Slavoj Žižek and including Giorgio Agamben and Alain Badiou—all thinkers not generally noted for being theologically inclined—have turned toward Christianity or, in the case of Derrida and Agamben, toward a Judaic-inflected notion of the messianic as a way of reconfiguring a certain strain of Marxist thought about what constitutes the political. Here I will focus on what are probably the two most diametrically opposed approaches to the politico-theological, both articulated through readings of Paul's epistles: on the one hand, Badiou's claim that Paul represents a model of revolutionary universalism, and on the other, Agamben's use of Paul's epistles to outline a theory of messianic time. It is toward a notion of what constitutes the messianic for Agamben and Badiou (and the latter does not embrace this term) that my observations are directed, and I will claim that there is a latent messianism embedded in Badiou's consistent preoccupation with questions of number.

It is not, I think, too strong a statement to assert that nearly all of Agamben's oeuvre is oriented toward demarcating a doubleness, whereby one thing is actually exposed to be two terms in relation, and the slight shift of perception that comes with this insight is, for Agamben, the mark of the messianic. In *The Coming Community*, Agamben delineates this mechanism very precisely, especially with his description of the virtually imperceptible shift that will take place with the advent of messianic time:

> There is a well-known parable about the kingdom of the Messiah that Walter Benjamin (who heard it from Gershom Scholem) recounted one evening to Ernst Bloch, who in turn transcribed it in *Spuren*: "A rabbi, a real cabalist, once said that in order to establish the reign of peace it is not necessary to destroy everything nor to begin a completely new world. It is sufficient to displace this cup or this bush or this stone just a little, and thus everything. But this small displacement is so difficult to achieve and its measure is so difficult to find that, with regard to the world, humans are incapable of it and it is necessary that the messiah come." Benjamin's version of the story goes like this: "The *Hassidim* tell a story about the world to come that says everything there will be just as it is here. Just as our room is now, so it will be in the world to come; where our baby sleeps now, there too it will sleep in the other world. And the clothes we wear in this world, those too we will wear there. Everything will be as it is now, just a little different."[1]

In the first recounting of the messianic story by Bloch, the marker of the slight displacement is something inhuman, a cup, bush, or stone, and the measure of the slight displacement so difficult to achieve that it is beyond the human. This situation of the messianic at the limit of the human perfectly anticipates Agamben's recent work, which is even more explicitly engaged with questions of life, the human, and the messianic.

In *The Open*, the encounter with the inhuman is exemplified by the animal, and it is a chiasmic opening to the closedness of the animal that is the marker of the human (though it is not clear that the animal need be present at all for this opening within the human to take place). Agamben outlines this dialectical relation of the open and the closed through a reading of Martin Heidegger, and he situates this encounter with the extrahuman space of blockage (which, again, is ultimately still part of the human) as the hallmark of the mystical:

Heidegger seems here to oscillate between two opposite poles, which in some ways recall the paradoxes of mystical knowledge—or rather, nonknowledge. On the one hand, captivation is a more spellbinding and intense openness than any kind of human knowledge; on the other, . . . it is closed in a total opacity. Animal captivation and the openness of the world thus seem related to one another as are negative and positive theology, and their relationship is as ambiguous as the one which simultaneously opposes and binds in a secret complicity the dark night of the mystic and the clarity of rational knowledge.[2]

For Agamben, then, the rational is inseparable from the mystical, the positive from the negative, life from death. This is not so much a redemptive reading of the "negative" term—though it is that too—as it is a focus on the complexity of relation between two contradictory movements that nonetheless reside together in the same place or entity, so much so that it would be easy not to perceive their distinct valences.

In *State of Exception*, Agamben links this mystical element to the state of exception and in so doing mentions in passing that it is not unlike Claude Lévi-Strauss's "floating signifier."[3] What is notable about this formulation—and crucial for the connection to Badiou that I will develop—is that the floating signifier is above all the index of a type of relation, one between signifier and signified that is marked by the excess of the signifier but is nonetheless inextricably linked to the structural relation between the two terms. As Gilles Deleuze emphasizes in his reading of Lévi-Strauss in "How Do We Recognize Structuralism?" the structure itself is more important than the set of (often fourfold or chiasmic) kinship or mythical relations it is evoked to analyze. Yet because it is at once the driving analytic motor and virtually devoid of signification in its own right, it is not unlike a mystical glue. Deleuze writes, "The Structure, consisting of the interrelation of the distinct entities, remains in itself opaque, thwarting all attempts at interpretation."[4] So, too, although Agamben never quite says this in so many words, it seems that his readings are as much about the structure of relation as they are about the concrete entities related. In *State of Exception*, he differentiates between two related yet distinct elements regarding the law. One is the normative and juridical element of law, and the other is the state of exception from law, what he also refers to as "force-of-law," where "what is at stake is a force of law without law."[5] "As long as the two elements remain correlated yet conceptually, temporally, and subjectively distinct,"

Agamben writes, ". . . their dialectic—though founded on a fiction—can nevertheless function in some way. But when they tend to coincide in a single person, when the state of exception, in which they are bound and blurred together, becomes the rule, then the juridico-political system transforms itself into a killing machine."[6] Here, the state of exception, the force-of-law, is the mystical glue that underpins the whole system, and when it is made to coincide entirely with the law, the dialectic of slight displacement that kept the system running breaks down. Indeed, it is when two parts of a system not perceived to be distinct are forcibly unified that the system faces its biggest crisis and potential for breakdown (which is precisely what Marx argues in "Crisis Theory"[7]). At issue, then, is the necessity of the messianic for preserving the disjunctive tension between two terms that might risk being collapsed into one.

Agamben's emphasis on the messianic and even the mystical would seem to put his book on the apostle Paul resolutely at odds with Badiou's book on Paul, which directly preceded it and which Agamben comments on obliquely yet critically. In many respects—argumentative, stylistic, and otherwise—these readings could not be more dissimilar. Yet where Badiou and Agamben converge, and I will return to this in conclusion, is in a preoccupation with the complexity of number and counting, with two terms that may be mistaken for one, and with the importance of registering this dialectic as a relation between two rather than a pure one. Before exploring this convergence—as well as the way these two seemingly opposed texts on Paul might share a secret preoccupation with messianic time—it is first necessary to situate and outline Badiou's reading of Paul within the context of his larger oeuvre.

In *Saint Paul: The Foundation of Universalism*, Paul serves as a model and mouthpiece for Badiou's philosophical system. As someone who remains faithful to the event of Christ's resurrection, founding through this a universal truth-procedure that shuns the particular, the dialectic, and the law, Paul is the exemplary militant figure for Badiou. One of the refrains of Badiou's analysis is the way in which Paul does away with the dialectic between life and death: "For Paul, there is an absolute disjunction between Christ's death and his resurrection. For death is an operation in the situation . . . while resurrection is the event as such."[8] Although Christ had to die in order to be resurrected, the event of the resurrection, which belongs to the domain of grace, is an effect that outlives and supersedes its cause, much as for Paul both the Greek cosmological order and the Jewish law are superseded by an overarching third term, Christianity. Badiou's reading of Paul shuns the mention of death in any fashion, even the dialectical relation of Christ's death as conduit to everlasting life. Unlike Agamben, whose writings repeatedly focus on this hinge point between life and death, Badiou's thought leaves no room for a space of ambiguity or relation between the two terms. If anything, Badiou the avowed atheist affirms this supersession by the third term in a less ambiguous fashion than Paul himself. This is particularly evident in Badiou's discussion of Paul's treatment of the relation between law and sin in his Epistle to the Romans. Unlike the event of the resurrection, the law is particular rather than universal and is in direct relation to desire and death.[9] For Badiou, law is akin to a type of evil that is not in itself a radical evil but has an evil function insofar as its prohibitory structure creates desire, which leads directly to sin and death.

Such a reading is in diametric opposition to Jacques Lacan's juxtaposition of Immanuel Kant and the Marquis de Sade in Lacan's seventh seminar, *The Ethics of Psychoanalysis*, and especially in the essay "Kant with Sade," published several years after the 1959–60 seminar.[10] Though the spirit of Kant's and Sade's works could not be more dissimilar, Lacan focuses rather perversely on a striking parallel of form, demonstrating that Kant's categorical imperative of acting so that one's maxim is always universalizable takes on the same structure as Sade's formula of always maximizing one's pleasure. Furthermore, both formulations are in fact based in form and not content, for they give no specific measures of conduct. In a particularly noteworthy twist at the end of "Kant with Sade," Lacan upbraids Sade, as it were, for not recognizing the structure of desire that is linked to the dialectic itself, to the dialectic of law: "Sade thus stopped, at the point where desire is knotted together with the law. If something in him held to the law, in order there to find the opportunity Saint Paul speaks of, to be sinful beyond measure, who would throw the first stone? But he went no further."[11] Here, Lacan points out that Sade, like Freud, holds too strictly to a *refusal* of the Christian precept of loving the neighbor to escape as he might want the structure of its law, which is in this case the law of selfless love that Paul—and Badiou following him—would see as not a law at all.[12]

Like those who for Paul are still beholden to Jewish law, Lacan's Sade remains entangled in a dialectic of law and desire. And Lacan would imply that the apostle Paul is himself not immune from such an entanglement.

Could it also be that Paul, in his discussion of sin and its relation to the law in the seventh chapter of his Epistle to the Romans, does not elaborate just what "sinful beyond measure" might mean (Romans 7:13)?[13] Just what is this "opportunity," as Lacan puts it, of being sinful beyond measure, and is there not some desire at work in Paul's letter that shows a drive toward that unrealized opportunity? Is there not a certain tension at the heart of Paul's letters, an indicator that, if nothing else, Paul was caught up in a certain dialectic of life and death, even though his message is of choosing life as a means beyond this very dialectic? In the process of hailing this life beyond death, Paul simultaneously imagines a "sin beyond measure" that in some scenario might be the very inhuman limit that is a consistent preoccupation of both Lacan and Agamben and that, according to Badiou, Paul never broaches.

Yet things seem a bit different if we look at the beginning of the seventh chapter of the book of Romans, which is none other than an example of the law concerning marriage. Paul uses the death of the husband as an analogy for the way that life in Christ supersedes the law:

> Thus a married woman is bound by the law to her husband as long as he lives; but if her husband dies, she is discharged from the law concerning the husband. Accordingly, she will be called an adulteress if she lives with another man while her husband is alive. But if her husband dies, she is free from that law, and if she marries another man, she is not an adulteress. In the same way, my brothers, you have died to the law through the body of Christ, so that you may belong to another, to him who has been raised from the dead in order that we may bear fruit for God. (Romans 7:2–5)[14]

It is odd that the example Paul employs to illustrate the merits of the law-superseding life in Christ takes as its pivotal point the death of the husband. For it is only through the husband's death that the law of marriage is suspended.[15] If anything, this passage from Romans affirms Antigone's unorthodox claim that a husband and children are ultimately replaceable but a brother is not. Is this not to some degree an example of the expendability of the couple and the permanence of the (unnatural?) family—here the brothers in Christ?[16] Not only, it would seem, does Paul envision, surely in spite of himself, a realm of sin beyond measure but also one of the impermanence of the couple. It is this heterodox Paul, the one who "went no further" than this, that Badiou views exclusively as a radical subverter and overturner of the law. This is certainly also the case, but following Lacan, I would suggest there is simultaneously a more perverse logic at work in Paul's epistles.

This perverse logic is one that is bound up with the dialectic of life and death, the same dialectic that is at the heart of Agamben's work. Interestingly, Agamben does not highlight the life-death dynamic in his book on Paul as much as in many of his other works, yet the book's central themes of the remainder and of messianic time have the same structural logic of this dynamic. Although in *The Time That Remains: A Commentary on the Letter to the Romans* Agamben explicitly devotes only two pages to a critique of Badiou's notion of universalism, the entirety of his philologically based close reading of Paul's epistles might be seen as pitted against the pure positivism of Badiou's reading, for Agamben repeatedly gestures to a richness and ambiguity in Paul's language that is at odds with a simple overturning of the law.

Agamben argues that the very act of wanting to overturn the law, especially insofar as the law can be linked to the state, is itself symptomatic of an inability to escape the thought structures of the law and the state. And this is precisely the point that Lacan makes in arguing that Sade ultimately reinforces the Kantian system rather than undermining it, for the very fact of trying to overturn the law or the state means that one is still embedded in its logic. As Agamben writes with respect to Badiou:

> This is how, in the book just referred to, Badiou is able to think about Paul's universalism as "benevolence with regard to customs and opinions" or as an "indifference that tolerates differences". . . . Despite the legitimacy of concepts such as "tolerance" or "benevolence," which in the end, pertain to the State's attitude toward religious conflict (one can see here how those who declare their wanting to abolish the state are often unable to liberate themselves from a point of view of the state), these concepts are certainly not messianic. For Paul, it is not a matter of "tolerating" or getting past differences in order to pinpoint a sameness or a universal lurking beyond. The universal is not a transcendent principle through which differences may be perceived—such a perspective of transcendence is not available to Paul. Rather, this "transcendental" involves an operation that divides the divisions of the law themselves and renders them inoperative, without ever reaching any final ground. No universal man, no Christian can be found in the

depths of the Jew or the Greek, neither as a principle nor as an end; all that is left is a remnant and the impossibility of the Jew or the Greek to coincide with himself. The messianic vocation separates every *klēsis* from itself, engendering a tension within itself, without ever providing it with some other identity; hence, Jew *as non-*Jew, Greek *as non-*Greek.[17]

This passage captures many of the central claims that Agamben makes about the messianic and correlatively about messianic time. Rather than insisting on the separation between the Christian and the Jew or the Christian and the Greek, Agamben tries to show that Paul does not merely divide Jew from non-Jew but also makes other divisions, such as that between spirit and flesh. What ensues in this fashion is a whole series of divisions such that there is always a remainder, always something that is not clearly divisible into one category or another. Agamben links this notion of the remainder to the way Christ's coming renders the old law inoperative. Indeed, this insistence on the positive force of the inoperative is central to Agamben's whole oeuvre, for the inoperative is not simply death but a force of suspension of life that might be said to enhance life by the very proximity it opens to the space of the negative. This force of suspension might be translated, to evoke a phenomenological resister that is not exactly Agamben's, as a structure of perception that is able to look unflinchingly upon seemingly negative states—death, stuckness, immobility, inertia—and take away from this encounter a potential for transformation. At issue is thus not simply the negative or death, nor is it the liminal space between life and death, but it is the perceptual encounter with that negative space from another space that is distinct from it. In this regard, Antonio Negri's critique of Agamben as simply reiterating a Heideggerian preoccupation with death seems like a wrong-spirited reduction of Agamben (and, for that matter, of Heidegger).[18]

Agamben, in fact, goes to great lengths to show the difference between overturning a law and making it inoperative, dwelling on Paul's repeated use of the verb *katargein*, which would indicate coming out of an act, or being suspended, as opposed to undoing or destroying, as this verb is often rendered. In this fashion, it is not so much a question of a dialectical undoing of the law but rather of its suspension or of its being rendered inoperative. For example, Agamben notes how the word *katargein* is formed from the word *argeō*, which in the Septuagint indicates rest on Saturday. Agamben writes, "It is certainly not by chance that the term used by the apostle to express the effect of the messianic on works of the law echoes a verb that signifies the sabbatical suspension of works" (*TTR*, 96).[19] By this reading, the sabbatical suspension is perfectly parallel to the state of exception. In other words, the sabbatical suspension is to the working days of the week what the force of law is to law: at once part of the entity in question, even a sort of end point toward which it is oriented, yet also that time or space in which the rules of the workweek or the law are overturned, effectively the exception that proves the rule. Agamben includes a short section just after this passage titled "State of Exception," in which he poses the question, "What is a law that is simultaneously suspended and fulfilled?" and goes on to respond by referring to Carl Schmitt and Jacob Taubes, linking their political theology to Pauline messianic suspension of law, an argument fully anticipating *State of Exception* (*TTR*, 104).

Agamben's interest in these questions is not unique to him, falling as it does at the end of a lineage of twentieth-century Judaic-inflected writings on the messianic and, more specifically, on messianic time. Agamben concludes *The Time That Remains* with a chapter on Benjamin and proposes several links between Benjamin's concept of *Jetztzeit*—the time of the now, the time that can "blast out of the continuum of history"—and its "*weak* messianic power," with the Pauline messianic.[20] The book's penultimate paragraph begins, "Whatever the case may be, there is no reason to doubt that these two fundamental messianic texts of our tradition [Paul and Benjamin], separated by almost two thousand years, both written in a situation of radical crisis, form a constellation whose time of legibility has finally come today, for reasons that invite further reflection" (*TTR*, 145).[21] Agamben's reflections on messianic time are preceded not only by Benjamin but even more proximately by Derrida's *Specters of Marx*, which arguably inaugurates the whole turn toward the politico-theological of the past decade. Through a reading of *Hamlet*, Derrida characterizes messianic time as a time "out of joint." Downplaying any links to an actual Judeo-Christian messianism, Derrida labels this time (or even space) a "messianicity without messianism."[22] Without further elaborating these examples, I would simply suggest that Benjamin, Derrida, and Agamben all draw on a certain Judaic vision of the messianic while simultaneously keeping it at arm's length.

Given the parameters of this genre of evoking the messianic but without recourse to messianism proper, the addition of a fourth thinker to this group of three takes on added significance. The thinker in question is the late Jacob Taubes, whose lectures collected as *The Political Theology of Paul* were given in Heidelberg in 1987 about *one month* before his death. (According to the preface to the posthumously published lectures, "Taubes, who did not have the strength to stand even for a moment, could lecture to us with the greatest intellectual intensity, *four* days out of the week, for *three* hours at a time, about his reading and contextualization of the Epistle to the Romans."[23]) What Taubes emphasizes, in several elliptical and startling moments in his lecture, is the fundamentally paradoxical if not perverse nature of Jewish messianism, and he does not shy away from embracing this term, something that neither Benjamin nor Derrida—nor ultimately even Agamben—embrace in such an unqualified fashion.

Taubes puts Paul's messianism squarely within a history not only of Jewish mysticism but also of Jewish apostasy: "This paradoxical faith is what I've tried to explain to you from the point of view of religious history with respect to the messianic logic in the history of Jewish mysticism, as a logic that is repeated in history. Whoever understands what Scholem presents in the eighth chapter of *Major Trends in Jewish Mysticism* can penetrate more deeply into Paul's messianic logic than by reading the entire exegetical literature."[24] This messianic logic touches on several dynamics at once: it recognizes the motor force of acts of transgression or apostasy; it "tarries with the negative" in that the perception it affords is accessible only through limit conditions such as death, illness, or despair; and it seeks to maintain a barely perceptible register of slight difference so that two things that might appear to be the same are recognized as distinct.[25]

Like the Lacan of the seminars, Taubes takes a continually provocative stance toward his listening audience, berating them for what they surely haven't read and do not understand and often not deigning to clarify for them the extraordinary connections and observations that are just thrown out and left there. Not surprisingly, then, Taubes does not say what it is that Scholem presents in the eighth chapter of *Major Trends in Jewish Mysticism*.

The eighth chapter, "Sabbatianism and Mystical Heresy," treats Sabbatai Zevi's relation to Nathan of Gaza, his young disciple who around 1665 (the year Spinoza suspended his work on the *Ethics* to write the *Theologico-Political Treatise*) had a Damascus-like revelation, except it was in Gaza. The revelation was that Sabbatai Zevi—who at that point, according to Scholem, was a rather average manic-depressive mystic—was, in fact, the Messiah.

According to Scholem, it was this revelation to Nathan of Gaza that in turn allowed Sabbatai Zevi to proclaim in Gaza that he was the Messiah. As Scholem writes:

> Nathan represents a most unusual combination of character traits. If the expression be permitted, he was at once the John the Baptist and the Paul of the new Messiah, surely a very remarkable figure. He had all the qualities which one misses in Sabbatai Zevi: tireless activity, originality of theological thought, and abundant productive power and literary ability. He proclaims the Messiah and blazes the trail for him, and at the same time he is by far the most influential theologian of the movement.[26]

Insofar as Nathan of Gaza is both John the Baptist and Paul to Sabbatai Zevi's Christ, he also embodies the out-of-jointness of messianic time, for he is at once the precursor and the follower of the Messiah.[27] This is an upending and dizzying form of time, not without its additional affinities to what Freud has termed *Nachtraglichkeit*, a kind of uncanny repetition, the trauma that is only experienced belatedly and retrospectively once it is finally repeated, but in a slightly new fashion.

The other aspect of Sabbatai Zevi that Scholem emphasizes is that he conjoins the figure of the Messiah with that of the sinner-apostate. Scholem highlights the paradox of this conjunction by comparing Sabbatai Zevi to Christ:

> To return to our comparison, the fate of the Messiahs is entirely different and so is the religious paradox. The paradox of crucifixion and that of apostasy are after all on two altogether different levels.... Death and apostasy cannot possibly evoke the same or similar sentiments, if only because the idea of betrayal contains even less that is positive. Unlike the death of Jesus, the decisive action (or rather, passion) of Sabbatai Zevi furnished no new revolutionary code of values. His betrayal merely destroyed the old. And so it becomes understandable why the deep fascination exercised by the conception of the helpless Messiah who hands himself over to the demons, if driven to its utmost limits, led directly to nihilistic consequences.[28]

This nihilistic vision of the messianic is what I take to be at the heart of Taubes's lectures on Paul at the end of his life. But here the term *nihilism* should be qualified, because what is at issue is not so much nihilism per se, but the way divine apostasy may pass for nihilism. As Slavoj Žižek points out in *The Puppet and the Dwarf*, Judas's betrayal, when read from a different perspective, might be seen as a higher form of fidelity in that it allows for the carrying out of Christ's divine mission.[29] It is also notable that Scholem in this passage contrasts the paradox of Sabbatai Zevi's apostasy to that of Christ's death, and while apostasy represents the more damnable state, it is also not simply reducible to death-as-the-negative—it is something that, in the spirit of the messianic, registers beyond the simple boundary of life and death. Apostasy is not equatable with death, nor is it purely the space of the negative; rather, it represents the ability to enter into that space, to be transformed by it, and ultimately to bear with it. Apostasy is more extreme in that it exemplifies the perverse logic that traverses the Jewish messianic moment. Indeed, it is the thought of the sin beyond measure.

This sin beyond measure seems proximate to what Taubes characterizes with reference to Benjamin as "creation as decay": "Benjamin—this is the astonishing parallel—has a Pauline notion of creation; he sees the labor pains of creation, the futility of creation. All of this is of course to be found in Romans 8: the groaning of the creature. Open this text and read it out loud, and then read Benjamin: You're going to be amazed. Romans 8:18. That's what Benjamin is talking about. That is the idea of creation as decay, since it is without hope."[30] What I want to distill from this is, once again, not so much an out-and-out nihilism or hopelessness, but the act and force of energy and disjointed temporality entailed in coming to that recognition, which is not without profound relation to the movement of the Pauline revelation but perhaps represents its darker side. Again, this is not nihilism, not eschatology, but a recognition of a difference of one, of one month, of one number in the count, often perceived at the end of a life.

This difference of one signals the importance of the count, of counting and numbers and dialectic as they relate to the politico-theological, which is ultimately what gives the work of Agamben and Badiou a certain unlikely proximity. Given the philological emphasis of Agamben's analysis, it is only a small surprise when it turns out that the very same verb, *katargein*, has a link to none other than the Hegelian dialectic. As Agamben writes, with no small element of provocation:

> At this point, I must return to the discovery I alluded to concerning the posthumous life of the verb *katargein* in the philosophical tradi-

tion. How does Luther translate this Pauline verb, whether in Romans 3:31 or wherever else the verb occurs in the Letters? Luther uses *Aufheben*—the very word that harbors the double meaning of abolishing and conserving (*aufbewahren* and *aufhören lassen*) used by Hegel as a foundation for his dialectic! . . . This is how a genuinely messianic term expressing the transformation of the law impacted by faith and announcement becomes a key term for the dialectic. That Hegel's dialectic is nothing more than a secularization of Christian theology comes as no surprise; however, more significant is the fact that (with a certain degree of irony) Hegel used a weapon against theology furnished by theology itself and that this weapon is genuinely messianic. (*TTR*, 99)

Agamben goes on to suggest that the Hegelian dialectic is nonetheless not messianic enough, for it foregrounds too emphatically the third and resolving term that would gesture toward a more apocalyptic end of history (and he connects this to the apocalyptic version of Hegel presented by two Franco-Russian Hegel scholars, Alexandre Koyré and Alexandre Kojève). Agamben is again at pains to distinguish the messianic, which represents a sort of suspended present time, from the apocalyptic or eschatological, which focus on the end. What Agamben does not envision—and this is where I will return to Badiou in a more redemptive fashion—is the possibility of a fourth term of the dialectical process that would capitalize on and in fact *be* the realm of the messianic that is already embedded in the *Aufhebung* of the Hegelian, and Marxian, dialectic.

Badiou notes at one point in *Saint Paul* that Hegel gestures to this fourth term at the end of his *Logic* (*Saint Paul*, 41). Badiou himself mentions at several points that perhaps there should be four terms in his analysis of Paul: the Greeks, the Jews, the Christians (which Paul represents), and the mystics. This fourth term is "the discourse of the ineffable, the discourse of nondiscourse" (*Saint Paul*, 51). Badiou goes on to elaborate:

> The fourth discourse (miraculous, or mystical) must remain *unaddressed*, which is to say that it cannot enter into the realm of preaching. . . . For Paul will not permit the Christian declaration to justify itself through the ineffable. . . . Let us say that, for Paul, the ethics of discourse consists in never suturing the third discourse (the public declaration of the Christ-event) to the fourth (the glorification of the subject personally visited by miracle). This ethics is profoundly coher-

> ent. Supposing I invoke . . . the fourth discourse . . . and hence the private, unutterable utterances, in order to justify the third (that of Christian faith), *I relapse inevitably into the second discourse*, that of the sign, the Jewish discourse. (*Saint Paul*, 52–53)

While acknowledging this fourth space, this is also the space that Badiou resoundingly avoids, leaving him in this regard not un-Hegelian. This fourth space is too close to the Judaic and, though Badiou himself does not use the term, too close to the state of suspension or exception that is the hallmark of Agamben's messianic.[31]

This refusal of the fourth term is consistent within the context of the book on Paul, but it is strikingly at odds with the larger corpus of Badiou's oeuvre, which is systematically organized around the four-part breakdown of what Badiou terms "generic procedures" into the domains of politics, art, science, and love. Badiou's concept of the generic procedure is in many ways hard to define, but examples of such a truth procedure might be the events of May 1968, the intellectual energy around French Maoism, or the model of love between two people. My point here is not so much to explicate the truth-event as Badiou outlines it but rather to note that it is singularly accessed through the four categories of politics, art, science, and love. These categories recur with exemplary consistency throughout Badiou's work.[32]

This penchant for four-part schemas has an interesting resonance with Badiou's multiple writings on the philosophy of mathematics and numbers. In a commentary on Lacan (who is also notably drawn to four-part schemas, as in *The Four Fundamental Concepts of Psychoanalysis* or the four discourses that he developed several years later in his seventeenth seminar, *The Other Side of Psychoanalysis*[33]), Badiou expressly addresses the differences between the numbers one, two, three, and four. In reference to a four-part system in Lacan, he asserts: "To attain 3, we need to *come back from* 4. But what is 4? 4 is the cipher of *discourse*. You know that any discourse organizes the $ of the subject, the S_1 of the master signifier, the S_2 of the derived signifier and the object *a*. The discourse is a figure of the 4. We can also say that with 4, we *know how to count*. We have, starting with 4, the *discourse of the Number*."[34] Now all this would reinforce the symbolic power of four as the ultimate structure of discourse and argumentation, something that Badiou's work as a whole certainly underscores. But what is striking here is the relation of three and four in Badiou's model. He notes,

"The 3 will be obtained by *descending from* 4, starting with the discourse. . . . The 3 has *skipped* between 2 and 4: *we can only obtain the structure starting with the discourse*. We can then only obtain the 3 starting from the 4."[35] What this typology of numbers represents is something actually quite close to the messianic structure that Agamben locates in Paul's epistles. Like the messianic, the three is suspended between the two and the four; it is the inoperative moment of the four of discourse. It is the four displaced, the subtraction of one. It is not unlike what Badiou writes about truth at the beginning of *Saint Paul*: "A truth is always, according to the dominant law of the count, subtracted from the count" (*Saint Paul*, 11).

This subtraction from the count is remarkably similar to what Agamben characterizes as the Saturday of messianic time. Agamben cites a rabbinic commentary on the book of Genesis proposing that Saturday represents the taking of an element of profane time and adding it to sacred time, but beyond that it stands in for the very displacement between Saturday and Sunday. Agamben comments, "Saturday—messianic time—is not another day, homogeneous to others; rather, it is that innermost disjointedness within time through which one may—by a hairsbreadth—grasp time and accomplish it" (*TTR*, 72). Insofar as the messianic can be represented by an adjacent day or number, which is not the last (that would be the apocalyptic) but the penultimate, not the seventh day as the day of rest but the sixth (the difference between the Christian and the Jewish Sabbath), and insofar as this space of difference in the count could be said to mark the messianic, then it seems that such a space of subtraction from the count is the latent messianic element in Badiou's work. More than subtraction in the sense of loss, it seems that Badiou uses this term to signal a difference of one, or a relation of the penultimate, as opposed to an exclusive recognition of only the last and most visible term. Why, then, in a work where Badiou makes the apostle Paul stand in as the marker of an ultimate universalizing truth-event—in other words, the epitome of Badiou's logical system—are there only three terms: the Jew, the Greek, and the Christian? Isn't the exulted category of the Christian based, as Agamben shows, on an inoperative and messianic fourth term from which the third category of the Christian might be read, not as the dialectical surpassing of the Greek and the Jew, but as itself the thing subtracted or displaced from its messianic remainder? If this were explicitly acknowledged, then the specifically Judaic dimension of Agamben's messianic might be opened—in a gesture partaking of apostasy or grace or both—to the Christian and even beyond.

Notes

1. Giorgio Agamben, *The Coming Community*, trans. Michael Hardt (Minneapolis: University of Minnesota Press, 1993), 53.
2. Giorgio Agamben, *The Open: Man and Animal*, trans. Kevin Attell (Stanford, CA: Stanford University Press, 2004), 59.
3. Giorgio Agamben, *State of Exception*, trans. Kevin Attell (Chicago: University of Chicago Press, 2005), 37.
4. Gilles Deleuze, "How Do We Recognize Structuralism?" in *Desert Islands and Other Texts, 1953–1974*, trans. Mike Taormina (New York: Semiotext(e), 2003), 37.
5. Agamben, *State of Exception*, 39.
6. Ibid., 86.
7. Karl Marx, "Crisis Theory," in *The Marx-Engels Reader*, ed. Robert C. Tucker (New York: W. W. Norton, 1978), 443–65.
8. Alain Badiou, *Saint Paul: The Foundation of Universalism*, trans. Ray Brassier (Stanford, CA: Stanford University Press, 2003), 70–71. Hereafter cited parenthetically by page number as *Saint Paul*.
9. Badiou links the law to the state and to death: "The law is always predicative, particular, and partial. Paul is perfectly aware of the law's unfailingly 'statist' character. By 'statist' I mean that which enumerates, names, and controls the parts of a situation. If a truth is to surge forth eventually, it must be nondenumerable, impredicable, uncontrollable. This is precisely what Paul calls grace" (*Saint Paul*, 76). He continues: "The law is what gives life to desire. But in so doing, it constrains the subject so that he wants to follow only the path of death. . . . The life of desire fixed and unleashed by the law is that which, decentered from the subject, accomplishes itself as unconscious automatism, with respect to which the involuntary subject is capable only of inventing death" (*Saint Paul*, 79).
10. Jacques Lacan, *The Ethics of Psychoanalysis, 1959–1960*, trans. Dennis Porter (New York: W. W. Norton, 1992). See also Jacques Lacan, "Kant with Sade," *October* 51 (Winter 1990): 55–75.
11. Lacan, "Kant with Sade," 74.
12. See Freud's damning critique of the commandment to "love thy neighbor as oneself" in *Civilization and its Discontents*, trans. James Strachey (New York: W. W. Norton, 1961), 62–67. For three essays that situate the love-hate relation to the neighbor in the register of the politico-theological, see Slavoj Žižek, Eric L. Santner, and Kenneth Reinhard, *The Neighbor: Three Inquiries in Political Theology* (Chicago: University of Chicago Press, 2005).
13. All citations from the Bible are taken from *The New Oxford Annotated Bible* (Oxford: Oxford University Press, 1994).
14. Instead of *brothers* here, the word *friends* is given in the Oxford version, but other versions render it as *brothers* or *brethren*, and the Oxford version notes that it is *brother* in the Greek.
15. For a fictional illustration of an extreme pushing to the limit of the laws of marriage, see Pierre Klossowski's trilogy, *Les lois de l'hospitalité* (*The Laws of Hospitality*) (Paris: Gallimard, 1965).
16. Jacob Taubes makes the observation that the real difference between the Old and New

Testaments is that the Old Testament recounts case after case of a barren woman conceiving, whereas human lineage and reproduction is, in the New Testament, entirely beside the point. See Taubes, *The Political Theology of Paul*, trans. Dana Hollander (Stanford, CA: Stanford University Press, 2004), 59. A more detailed discussion of this text will follow.

17 Giorgio Agamben, *The Time That Remains: A Commentary on the Letter to the Romans*, trans. Patricia Dailey (Stanford, CA: Stanford University Press, 2005), 52–53. Hereafter cited parenthetically by page number as *TTR*.

18 Cesare Casarino and Antonio Negri, "It's a Powerful Life: A Conversation on Contemporary Philosophy," *Cultural Critique* 57 (Spring 2004): 151–83, esp. 171–74.

19 Agamben, in turn, links the Greek word to a transliteration from the Hebrew that I have been unable to verify to my satisfaction. Other scholars have questioned some of Agamben's philological claims in this text, but such difficulties do not have direct bearing on my argument.

20 Walter Benjamin, "Theses on the Philosophy of History," in *Illuminations*, trans. Harry Zohn (New York: Schocken Books, 1969), 254, 261.

21 See also the chapter "Threshold or *Tornada*," in *TTR*, 138–45.

22 Jacques Derrida, *Specters of Marx: The State of the Debt, the Work of Mourning, and the New International*, trans. Peggy Kamuf (New York: Routledge, 1994).

23 Taubes, *The Political Theology of Paul*, xiii, my emphasis.

24 Ibid., 49–50.

25 An example of the latter is the celebrated distinction that Agamben makes between bare life (*zoē*) and form-of-life (*bios*), a difference that becomes remarkable only when extreme circumstances, such as that of the concentration camps, collapse the two into one. Agamben has been criticized for glorifying the negative and elevating bare life to the sublime. See Giorgio Agamben, *Homo Sacer: Sovereign Power and Bare Life*, trans. Daniel Heller-Roazen (Stanford, CA: Stanford University Press, 1998). Criticisms include Dominick LaCapra, *History in Transit: Experience, Identity, Critical Theory* (Ithaca, NY: Cornell University Press, 2004), and Michael Hardt and Antonio Negri, *Empire* (Cambridge, MA: Harvard University Press, 2000), 366.

26 Gershom G. Scholem, *Major Trends in Jewish Mysticism* (New York: Schocken Books, 1946), 295–96.

27 This is not unlike what Deleuze in *The Logic of Sense* characterizes as *Aion*, the time of becoming that conjoins both past and future, as opposed to *Chronos*, the more static time of the present. See Gilles Deleuze, *The Logic of Sense*, trans. Mark Lester and Charles Stivale (New York: Columbia University Press, 1990). Moreover, at the end of *The Time That Remains*, in trying to establish Paul as a secret inspiration for Benjamin, Agamben draws on Scholem's text on Sabbatai Zevi:

> Scholem's attitude toward Paul, an author he knew well and once characterized as "the most outstanding example known to us of a revolutionary Jewish mystic", is certainly not lacking in ambiguity. Yet the discovery of a Pauline inspiration [for Benjamin] must have bothered [Scholem], although he certainly never would have raised the issue himself. Nevertheless, in one of his books—just as cautiously as when, in a book on Sabbatai Sevi, he establishes a relation between Paul and Nathan

of Gaza—he seems to actually suggest, albeit in a cryptic fashion, that Benjamin may have identified with Paul. (*TTR*, 144)

Scholem's quote here is from *On the Kabbalah and Its Symbolism*, trans. Ralph Manheim (New York: Schocken Books, 1965).

28 Scholem, *Major Trends in Jewish Mysticism*, 308–9.
29 Slavoj Žižek, *The Puppet and the Dwarf: The Perverse Core of Christianity* (Cambridge, MA: MIT Press, 2003), 17. Such an argument is given added weight by the recent discovery of the redemptive scrolls relating to Judas.
30 Taubes, *The Political Theology of Paul*, 72. This essay has benefited from a very perceptive and critical anonymous reader. Against Taubes, this reader argues that, to the extent Benjamin addresses creation, it is with respect to the destructive force, and this is ultimately one that preserves hope, such as the force of divine violence. While I agree with this reading of Benjamin, I do not think that Taubes's recognition of hopelessness is itself without hope. That is, to acknowledge the overall hopelessness of a situation is not necessarily the position of despair it would seem to be. Such overall acknowledgment and recognition may be what makes the situation bearable from moment to moment.
31 See *Saint Paul*, 41, 52, 98–99, where Badiou at once makes an overture to the fourth term, the mystical, and just as quickly shuts it down. Badiou's relation to the Judaic is to this reader highly fraught, partaking, if it can be called this, of the philo-Semitic anti-Semitism that marks a text like Sartre's *Anti-Semite and Jew: An Exploration on the Etiology of Hate*, trans. George J. Becker (New York: Schocken Books, 1948). Exemplary in this regard is Badiou's recent collection of essays about the word *Jew*, *Circonstances, 3: Portées du mot 'juif'* (Paris: Lignes, 2005), translated as "Uses of the Word 'Jew'," in Badiou, *Polemics*, trans. Steve Corcoran (London: Verso, 2006), in which he goes so far as to claim, "It was above all the Nazis who, before anyone else, and with a rare zeal for following through, drew all the consequences from making the signifier 'Jewish' into a radical exception" (163–64). For a scathing review of this book that would see nothing redemptive in Badiou's strategy of subsuming the particularity of the Jew to a universal term outside the realm of victimization, see Éric Marty, "Alain Badiou: L'avenir d'une négation" ("Alain Badiou: Future of a Negation"), *Les temps modernes* 635–636 (November–December 2005/January 2006): 22–57.
32 For a critical analysis of Badiou's four terms of the truth-process and his notion of the event as it relates to ethics, see my "Why the Family Is Beautiful (Lacan against Badiou)," *Diacritics* 32:3–4 (2002): 135–51.
33 Jacques Lacan, *The Four Fundamental Concepts of Psychoanalysis*, trans. Alan Sheridan (New York: W. W. Norton, 1981), and *The Other Side of Psychoanalysis*, trans. Russell Grigg (New York: W. W. Norton, 2006).
34 Alain Badiou, "Logic and Philosophy" (or "One, Two, Three, Four, and Zero Too"), talk given at University of California, Irvine, April 11, 2002.
35 Ibid.

Penelope Deutscher

The Inversion of Exceptionality:
Foucault, Agamben, and "Reproductive Rights"

> The condition of the living being about to be annihilated by death resembles the state in which we find ourselves in the maternal womb, or in the state of vegetation.
> —Xavier Bichat, *Recherches physiologiques sur la vie et la mort*
>
> Just as in the fetus organic life begins before that of animal life, so in getting old and dying it survives its animal death.
> —Giorgio Agamben, discussing Bichat in *Remnants of Auschwitz*

Since it has not to date arisen as a question, is it possible to open a debate with Giorgio Agamben concerning the role of women's bodies in the politicization of life? What different inflections of life and of politicized life would result from an intermittent insertion "born of women's bodies"?

Michel Foucault suggests that new sites of intensification and problematization occurring toward the end of the eighteenth century included reproduction and the birth rate in addition to the mortality rate.[1] When he described this specific modality for seizing hold of life, Foucault's work opened itself up to a possibility that was little developed in his own work: an interrogation of the intersection between an eventual notion of "reproductive rights" and the constitution of reproductivity as a biopolitical substance, inflecting state-based and other attempts to suppress abortion and the concurrent resistance to those attempts.

Foucault discusses abortion rights on at least two occasions. On one, it arises as part of his discussion of the relationship between power and resistance. That which makes power "strong," such as the investment in the body, is prone to become a focus for counterclaims. It is in such terms that Foucault analyzes instances of state or medical panic at demands for free abortion. Respecting the "complex phenomena" of power, Foucault describes power's investment in the body as stimulating desire with respect to one's own body.

> But once power produces this effect, there inevitably emerge the responding [*dans la ligne même de ses conquêts, emerge inévitablement*] claims and affirmations [*la revendication*], those of one's own body against power, of health against the economic system, of pleasure against the moral norms of sexuality, marriage, decency. Suddenly, what had made power strong [*ce par quoi le pouvoir était fort*] becomes the means by which it is attacked. Power, after investing itself in the body [*Le pouvoir s'est avancé dans le corps*], finds itself exposed to a counter-attack in that same body. Do you recall the panic of the institutions of the social body, the doctors and the politicians, at the idea of non-legalized cohabitation [*l'union libre*] or abortion?[2]

That revolt and liberation are incited by, rather than external to, the investments of power needn't make, for example, the abortion activist more wary, and yet it also needn't preclude an awareness of the tactics by which one may consolidate what one means to oppose. The issue arises again in this interview:

> [Bernard-Henri Lévi]: This idea that . . . to be happy we must have sexual liberation is held basically by sexologists, doctors, and those who police sex [*policiers du sexe*]. . . .
> Foucault: Yes, and that is why they present to us a formidable trap. What they are saying, roughly, is this: "You have a sexuality; this sexuality is both frustrated and mute; hypocritical prohibitions are repressing it. So, come to us, tell us, show us all that, confide in us your unhappy secrets. . . ."

This type of discourse is, indeed, a formidable tool of control and power. As always, it uses what people say, feel, and hope for. It exploits their temptation to believe that to be happy, it is enough to cross the threshold of discourse and to remove a few prohibitions. But in fact it ends up repressing [*rabattre*] and dispersing [*quadriller*] movements of revolt and liberation . . .
BHL: Hence the misunderstanding of certain commentators: "According to Foucault, the repression or the liberation of sex amounts to the same thing." Or again: "Le Mouvement pour la liberté de l'avortement et de la contraception [a radical pro-choice group] and Laissez-les vivre [a pro-life group] are reducible to basically the same discourse [*c'est au fond le même discours*]. . . ."[3]

Agamben's response to Foucault is well known for its reluctance to designate as only a modern phenomenon the seizing-hold of life by the political. While he does not engage in a project intended to flatten the differences between the ways that life has been politicized, one of the doubtless unintended by-products of his response to Foucault has been a nonengagement with the place of women in the biopoliticization of life. The question that was more centered within Foucault's line of vision given his focus on, for example, nineteenth-century natalist politics, was that of the hystericization and preoccupation with women's bodies. Once women's reproductivity in matters of population control and of reproduction incitement becomes the political focus, in addition to an interest in the "quality" of neonatal life and the status of life more generally, no particular theoretical conflict is likely to arise from extending one's analysis to the study of legislative intervention into state regulation of matters of contraception and abortion. Though it is possible to speculate about why Foucault contributed only minor remarks on the topic, given his support for contemporary abortion rights and his small collaboration with the Groupe de l'information sur la santé on this point in 1973 when abortion was still illegal in France,[4] a Foucauldian analysis of formations surrounding abortion control and abortion rights is perfectly consistent with a Foucauldian biopolitics. It is not clear, however, that Agamben's approach lends itself easily to a similar reflection.

To be sure, Agamben's silence on the possible question could be one of a theorist's many "nonaccidents." Some reasons would be theoretical. The examples of bare life he considers are usually formations that one can imagine having been identified as human life and then stripped of that status or subjected to a threshold status: the overcoma, the immigrant, the refugee, the internee, the enemy combatant, the *Muselmann*. A consideration of fetal life does not fit the series, as it usually is not situated at the threshold of depoliticization or dehumanization of previously politicized or humanized life. The fetus represents the zone of contested and intensified political stakes around the threshold between what some would consider "prelife" and what is to be identified as nascent human life, meaningful human life, and/or rights-bearing life.

Thus the ambiguous politicized life least separable from some women's bodies happens to be a formation least appropriate for Agamben's analysis. An emergent fetus usually is not considered to have had a political, legal, or linguistic status subsequently suspended. Rather, its original ambiguity is in contention when it comes to the anxieties of biopolitics. This may be one reason fetal life, despite being one of the major nodes of biopolitics, makes only the faintest of appearances in Agamben's work, remaining an ambiguous threshold of life in which he has been least interested.

Yet there may also be strategic reasons for Agamben's lack of interest in the biopolitics of women's reproductivity.[5] Once the interconnections between biopower and women's reproductivity are considered, how can one avoid an engagement with the history of abortion regulation? The intense interrelation between biopolitics and abortion legislation overlaps with Foucault's "threshold of modernity."[6] Once that interconnection is considered, one is confronted with the complex definitions of fetal life that, in fact, Agamben touches on in his discussion of Xavier Bichat in *Remnants of Auschwitz*. What happens when dying life is likened to fetal life, to the threshold between animal life and organic life, perhaps to bare life? Though his attention is directed more toward the threshold between human "being" and bare or barely "human," and thus in some respects more toward the threshold relevant to thanatopolitics rather than prenatal politics, Agamben makes occasional references to the excess of the human to the human being according to both forms of threshold, seen more than once in the formulation "beyond or before": "The human being is thus always beyond or before the human, the central threshold through which pass currents of the human and the inhuman, subjectification and desubjectification, the living being's becoming speaking and the logos' becoming living."[7] Yet this question of the before and the question of the prenatal, more specifically, might have the potential to trouble Agamben's concerns.

Consider the form of Andrew Norris's question in his introduction to a collection of essays on Agamben's *Homo Sacer*: "What, for instance, are we to do when we are dealing with agents or things that have not already been recognized as the bearers of rights?"[8] Once the question is posed in terms of that which "has not already been recognized" rather than that which is "no longer" recognized as the bearer of rights, what is at stake alters. Imagine that same question asked in a different context, at demonstrations taking place outside abortion clinics in the United States and elsewhere. Agamben's work might appear one step closer to an interested reading by the antiabortion activist whose extremism has extended to the passion for comparisons with Auschwitz.

This is a ghostly proximity not likely to take concrete shape in Agamben's work, partly because when he does consider those thresholds between "life and death, animate and inanimate, human and inhuman, nature and culture, law and bodies,"[9] his tendency is also to stress a consideration of a "new living dead man, a new sacred man"[10] and not the production of the threshold "prelife" or "prior to human life." However, it turns up intermittently in his references whenever he runs the post- and the prior together.

With the exception of Karen Quinlan, women's bodies are impressively absent from Agamben's writing, as are reproductive bodies. Yet it is true that the incorporation would not be a simple one, to such an extent that one of the conditions of his project could be located in the dissociation of "life" from "women's reproductivity." For the intriguing potential it has to operate as a lens to rethink the terms *life*, *bare life*, *threshold*, and *biopolitics*, it should be explored: Agamben's texts reappear as enveloping an alternative problematics that both unsettles and reorients these terms.

States of Exception

The biopolitical intensification and production of the ambiguous zone of the "prior to life" has a spectral and inverted relationship to Agamben's analyses. Not in order to create confusion with Agamben's, Carl Schmitt's, and Walter Benjamin's reflections on the state of exception, consider, for example, the inverted relationship between Agamben's state of exception and the states of exception associated with the history of abortion law in Europe, Britain, and Australia since the second half of the nineteenth century. Through its own and particular configuration, abortion has relentlessly and internationally—and in an uncannily duplicating formation of policy and law—been its own state of exception. Exceptionality has provided the form through which these countries have accomplished the legal and political regulation of abortion.

This form of exceptionality is not the state of exception Agamben discusses in the context of Schmitt and Benjamin—it is not that a nation's laws are set aside through such pretexts as a state of emergency. Instead, a particular practice—in this case, regular and legal abortion—has often taken shape through the granting of a general exception to an ongoing law, which, in fact (except for the exception), continues to render it illegal. This would not be an instance of defining sovereignty by the capacity to suspend the legal regime. But if such a capacity defines sovereignty (as with Schmitt's analysis of the state of exception), we could ask what kind of sovereignty, if any, is ensured or produced through the intertwining of reproductive practice with the laws organizing and criminalizing it, and allowing for broad exceptions from that criminality, sometimes lasting decades in a legal system. Thus the exception becomes regularized and regulative. It is not an entire regime that is the exception to its own illegality, but the laws addressing one phenomenon, abortion. The state of exception is not the state, not the nation or a country's suspended legal system; rather, it is abortion "itself" that has frequently existed in a state of suspension or exception to its own illegality.

An analytic of exceptionality would be required to understand the peculiar form of biopolitics covering women's reproductivity, the interest in termination, and the conflicting and agitated interests in naming what is termed "life," "rights-bearing life," or life requiring responses identified as "protection," which have taken extreme and, on occasion, murderous forms. The specificity of the form of exceptionality applied to this phenomenon is striking, as is the specificity of the states of exception discussed by Agamben. Do these overlap in any way? At first glance, it seems the answer must be no. When Agamben stresses the pretexts and conditions associated with states of emergency and the temporary suspension of a legal regime, our attention is directed to the sovereign's capacity to effect itself by setting aside almost an entire constitution (as in the passing of the Enabling Act in Germany in 1933, legally authorizing—with two conditions, dictatorial powers and the capacity to effect laws without parliamentary involvement—government measures that could "deviate from the constitution")[11] or its legislative regime (as in Raymond Poincaré's declaring France to be under a state of siege from 1914 to 1918, transferring legislative power to the executive).[12] But in Agamben's discussion, states of exception have also included suspension of particular national and international laws, as in the U.S. military order of November 13, 2001, which effectively suspended the applicability of the Geneva Conventions and existing American laws covering detained prisoners by denying to suspected so-called enemy combatants detained in the United States and Guantánamo Bay the status of foreign prisoners of war or the status of American prisoners. As Agamben writes, the possibility of indefinite detention—also discussed by Judith Butler as a suspension of the applicability of those international and national laws that otherwise would have pertained to the detainees[13]—was accomplished through the constitution of a new category:

> What is new about President Bush's order is that it radically erases any legal status of the individual, thus producing a legally unnamable and unclassifiable being. Not only do the Taliban captured in Afghanistan not enjoy the status of POWs as defined by the Geneva Convention, they do not even have the status of persons charged with a crime according to American laws. Neither prisoners nor persons accused, but simply "detainees," they are the object of a pure de facto rule.[14]

The USA Patriot Act (October 26, 2001) also operates through exceptionality.[15] It grants a blanket exception to liberties otherwise covered under the First, Fourth, Fifth, Eighth, and Fourteenth Amendments, which may be suspended in the "exceptional" circumstances of suspicion of terrorist activity. The constitutionality of the amendments has survived the enactment of the Patriot Act, and thus the act has granted an exception to amendments with which it would otherwise conflict. The wording of the Patriot Act was given maximum flexibility by inclusion of the capacious category "other purposes": it states that it is meant to "deter and punish American terrorists in the United States and around the world, to enhance law enforcement investigatory tools, and for other purposes." Exempting investigators from demonstrating probable cause and including the wording "other purposes," the scope of the act to suspend the terms of constitutional amendments is potentially broader than the ongoing "legality" of the latter.[16] The ambit of the enveloping exceptions could exceed the "ongoing" law, whose legality they don't ostensibly jeopardize.

Thus, the mode of exceptionality characteristic of some formulations of abortion law is not entirely unlike some forms of suspension discussed by Agamben. Particular laws are set aside within a legal regime and nevertheless persist within it, suspended to varying degrees. Each of these may be considered new exercises, forms, and constructions of sovereignty. Accordingly it is possible to ask what kind of sovereignty is effected through regularized but constantly volatile states of exception pertaining to legal regimes targeting women's reproductivity. Like the November 13, 2001, military order, these states of exception institute the fragility and centrality of bodies—reproductive no less than incarcerated. Whether the exception seems to protect while concurrently stressing the vulnerability of women's reproductive autonomy, or whether it seems to defend a state while weakening civil liberties, bodies are being intensified, weakened, and invested with their possible exposure to violence.

The Exceptionality of Abortion

From the 1820s through the rest of the nineteenth century in America, abortion past the fourth month of pregnancy was increasingly banned by individual states. Though this was reinforced in 1873 by the passage of the Comstock Law "for the Suppression of Trade in, and Circulation of, Obscene Literature and Articles for Immoral Use," which was applied to bans on obscene literature, information about birth control, and the practice of abortion, abortion remained a matter of state rather than federal law, having become illegal in all fifty states by the 1960s until the precedent established by *Roe v. Wade* in 1973. This case reconfigured abortion as included within the right to privacy protected under the Fourteenth Amendment of the U.S. Constitution:

> State criminal abortion laws, like those involved here, that except from criminality only a life-saving procedure on the mother's behalf without regard to the stage of her pregnancy and other interests involved violate the Due Process Clause of the Fourteenth Amendment, which protects against state action the right to privacy, including a woman's qualified right to terminate her pregnancy.[17]

As is clear in the ruling, these legal configurations have always been entangled with the language of exception. Throughout the nineteenth and twentieth centuries, the increasingly entrenched state-based criminalization of abortion usually allowed for exceptions under grounds such as rape or concern for the woman's life, health, or well-being. The bans, therefore, included exceptions that could, according to the contingencies of individual states, doctors, judges, contexts, and cases, allow for extreme variation in the actual liberality of access to abortion. And though *Roe v. Wade* is widely considered to have decriminalized abortion, Mary Poovey has noted how it simultaneously reconfirms the state's readiness to intervene, given its wording: "a woman's right to terminate her pregnancy is not absolute, and may to some extent be limited by the state's legitimate interests in safeguarding women's health, in maintaining proper medical standards, and in protecting potential human life."[18] Even the passage from the *Roe v. Wade* ruling, spelling out the new inclusion of abortion under the right to privacy, continues, "though the State cannot override that right, it has legitimate interests in protecting both the pregnant woman's health and the potentiality of human life, each of which interests grows and reaches a 'compelling' point at various stages of the woman's approach to term."[19]

In many countries also considered to have long decriminalized abortion, including France, Canada, Britain, and Australia, abortion's legal treatment has followed a different and yet consistently echoing pattern of establishing conditions for dispensation, exemption, or exception from an already established and continuing illegality without actual repeal of earlier laws rendering abortion illegal. For example, abortion in Germany was outlawed in 1871, an illegality that persisted under the Weimar Constitution, continued through the Third Reich (though it was intermittently forced on non-Aryan women), and is still technically illegal. While the current German law does not allow abortions, it suspends their legal prosecution under certain stated conditions: if the abortion takes place during the first twelve weeks of pregnancy and in conjunction with appropriate—pro-life-inflected—counseling; during the first twenty-two weeks of pregnancy on grounds deemed eugenic or criminological; or thereafter only for reasons considered medical. A large number of countries that have decriminalized abortion in the twentieth century have done so according to variations on this model.[20]

The form of exceptionality in question is one that has been analyzed by Michèle Le Doeuff in relation to France. Le Doeuff argues that the so-called Veil law, which was thought by many to have accomplished the decriminalization of abortion in 1975, acted as a reconfirmation of abortion's criminalization. It amounted to an exception reconfirming that abortion is illegal except under certain circumstances, however broad the scope of the latter may be. To enlarge on her example, in some contexts, as in Germany, where the exception effectively (though on diverse grounds) can apply throughout the entire term of the pregnancy, it is at least hypothetically possible that every abortion could be allowed, while every abortion remains nonetheless an exception to its own illegality. No matter how broad the set of exceptions, and even if the exception has become the norm and covers the entire field, the exception—this is the interpretation proposed by Le Doeuff—does reconfirm the illegality.

Under the French legal regime Le Doeuff considers, abortion was made illegal under the 1810 French Penal Code and was redeclared punishable in 1920 by the *cour d'assises*. According to a 1920 law, article 317 of the Penal Code, abortion was an offense, as was the dissemination of information about abortion and contraceptive products. Le Doeuff described subsequent modifications to the law as follows:

> The Neuwirth law, which permits the prescription of contraceptives and their sale in chemists, and the Veil law, which permits abortion in certain circumstances, are simply dispensations [*dérogations*] in relation to the law of 1920. . . .
>
> We should also recall here the correct formulation of an old legal adage: "The exception proves the rule *for non-excepted cases*." From this point of view, the dispensations provided by the Neuwirth and Veil laws correspond to a *reproclamation* of the 1920 law. . . . The legalization of an exception amounts to letting go of one element in order to uphold the fundamental point.[21]

Indeed, the phenomenon that preoccupies Le Doeuff could be thought of as the inverted form of an *Ausnahmezustand*: a particular kind of law, within a legal regime, that takes the relation to itself of setting itself aside, sometimes in certain specified conditions, and sometimes through the subsequent blanket or close to blanket exception to itself. This form of legality and exceptionality has been a primary form of incitement of, investment in, stimulation of, production of, and regulation of women's bodies as reproductive biopolitical targets. Whether or not women have had relatively easy access to abortion, the law deems this access either tenuous or an exception, however broad, and regulates the reinscribed possibility that women might not have that access and that they are enmeshed in a web of agitated interests. Arguably, forms of sovereignty relating to women can be located in these formations.

Agamben has argued that the state of exception is identified both as a ubiquitous technique of government but also as a "constitutive paradigm of the juridical order."[22] It could similarly be said of women's reproductivity that one identifies a parallel subregime that prefers the creations of permanent states of exception, and that the repeated creation of abortion as a state of permanent exceptionality has been one of the essential workings of twentieth- and early-twenty-first-century biopolitics concerning women's reproductivity. Agamben's inflection draws attention not just to the technique but also to the peculiarity of a constitutive paradigm. Perhaps there is room for those interested in the intersection of biopolitics and women's reproductivity from the nineteenth century to think further about the significance of the forms, juridical and otherwise, through which women's reproductivity is produced as a concurrent target and result of that form. Agamben notes: "There are not *first* life as a natural biological given and anomie as the state of nature, and *then* their implication in law through the state of exception. On the contrary, the very possibility of distinguishing life and law, anomie and *nomos*, coincides with their articulation in the biopolitical machine."[23] There are parallels between Agamben's statement and the phenomena of all targets of reproductive biopolitics and their productions: languages, preoccupations and effects of freedom, life, fetal life, personhood, potential personhood, right to life, rights over one's body, autonomy, and privacy.

One could also ask around what kinds of bodies one sees this preference for continuing a law in relation to which a modality of sovereignty is pursued in the granting of dispensation from its own harshest version, such that the exemption both suspends and reconfirms the harshest rule. Examples might be located in the practices of pardon, long-term death row incarceration, and sentence commutation, all constituting fields of exception to a death penalty regime whose legality is not weakened, and in exceptions regularly granted to national immigration restrictions. This is the case with exceptional but not infrequent regularizations of illegal or "undocumented" immigrants in France, the United States, and Australia, occurring under the guise of exceptions to highly restrictive immigration laws that, if anything, can be maintained in their harshest versions under these circumstances.

This would be the inverted version of the state of emergency regarding civil liberties. Obviously, the exceptions can constitute a means of reconfirming, in an intensified version, the regime of regulation. To return to abortion, in contexts enamored of this particular structure of exceptionality, there is almost never an illegal abortion, while it's just as true that there is almost never a legal abortion.

Homo Sacer

What consequences would arise from an interrogation of the politicization of women's reproductivity from the perspective of Agamben's work? Might some light be shed on the treacherous nature of the rhetoric—not within Agamben's work but in antiabortion contexts—concerned with "potential life"? This is the parallel that leads one to not include abortion in the context of Agamben's discussion of threshold life, in addition to the point that this is not a life whose humanity has been stripped or lost. If it has any temporality at all, it would be the temporality of the prior, not the post-.

In an uncanny ghosting, it is not uncommon for extreme antiabortion activists to make connections between abortion and extermination, invoking the specter of the Holocaust. Antiabortionists have been highly attuned to ruses for the politicization of the threshold of life, greatly attached to an appropriation of a category through the rhetorical or conceptual construction of fetal life as what one is tempted to think of as the pseudo–*homo sacer*. If the fetus is deemed akin to the subject in the camp, the antiabortionist figures the woman's body as a kind of camp. Such a camp is not unlike the detention camp of the illegal immigrant about whom the law adjudicates whether his or her life is to be considered a rights-bearing life. If so, it has often been noted that the rights—in cases in which they are conceptualized as such—are considered to make a kind of competing claim on the woman, as if they make competing rights claims on each other. Thus the woman seems to be slyly attributed the status of sinister sovereign, at the mercy of whom the fetus exists in its threshold state. It is also noteworthy that the woman's possible sovereignty may be considered a zone of disputed authority with an alternative sovereign power, the state. Recall the wording of *Roe v. Wade* that approximates the rhetoric of competing sovereign interests: "Though the State cannot override that right, it has legitimate interests in protecting both the pregnant woman's health and the potentiality of human life, each of which interests grows and reaches a 'compelling' point at various stages of the woman's approach to term."[24]

The woman legally forbidden to have an abortion is sometimes figured as a potentially murderous competing sovereign whose self-interest would thwart the intervening motivations of the state concerned with the threshold life in question. The alternative intention with which she is attributed is a pseudoviolent decision that this fetal life is not to be lived. Neither *zoē*, *bios*, bare life, nor *homo sacer*, the fetus is rhetorically and varyingly depicted as all of these, in an imitation of these patterns as they take place around *zoē*, *bios*, and the production of bare human life as vulnerable excess to which political life can be reduced. Agamben's analyses illuminate the way in which fetal life can come to be considered, particularly in antiabortion contexts and erroneously as a form of politicized bare life exposed to sovereign violence.

In the production of fetal life as a pseudo–*homo sacer*, we can usefully ask what has happened to the woman's body. Agamben's reflections have not encompassed an engagement with women's reproductive bodies. Yet one embarkation point for a potential exploration is to be found in his mention of "the woman" as one of the social-juridical entities superceding "the Marxist scission between man and citizen," and as such grouped with the series he proposes, which includes the voter, the worker, and the transvestite: "The Marxian scission between man and citizen is thus superseded by the division between naked life (ultimate and opaque bearer of sovereignty) and the multifarious forms of life abstractly recodifed as social-juridical entities (the voter, the worker, the journalist, the student, but also the HIV-positive, the transvestite, the porno star, the elderly, the parent, the woman) that all rest on naked life."[25]

It is surely fair to name the woman's reproductive body that which Agamben would prefer not to mention in his considerations of life. So we can and should ask, who is the *woman* who as a social-juridical entity rests on its division from naked life? My suggestion is that it is a woman whose status as potentially reducible to naked life is associated with her reducibility to reproductive life. This is the paradox of figuring the woman as a threatening and competing sovereign power over the fetus that is falsely figured as *homo sacer*: to do so is simultaneously to reduce the woman to a barer, reproductive life exposed to the state's hegemonic intervention as it overrides the woman erroneously figured as a "competing sovereign" exposing life. As she is figured as that which exposes another life, she is herself gripped, exposed, and reduced to barer life.

Notes

1 Michel Foucault, *"Society Must Be Defended": Lectures at the Collège de France, 1975–76*, ed. Mauro Bertani and Alessandro Fontana, trans. David Macey (New York: Picador, 2003), 243–44.
2 Foucault cautions, "The impression that power weakens and vacillates here is in fact mistaken; power can retreat here, re-organize its forces, invest itself elsewhere . . . and

so the battle continues." Michel Foucault, "Body/Power," in *Power/Knowledge: Selected Interviews and Other Writings, 1972–1977*, ed. Colin Gordon (New York: Pantheon, 1980), 56. Translation modified.

3 Michel Foucault, "Power and Sex," in *Politics, Philosophy, Culture: Interviews and Other Writings, 1977–1984*, ed. Lawrence D. Kritzman, trans. Alan Sheridan (New York: Routledge, 1988), 114. Translation modified.

4 One can consult Michel Foucault's 1973 piece, written in collaboration with Alain Landau and Jean-Yves Petit for the *Nouvel Observateur*, "Convoqués à la PJ," in Foucault, *Dits et écrits, 1954–1988*, vol. 1, *1954–1975*, ed. Daniel Defert and François Ewald (Paris: Gallimard, 2001), 1313–15. Foucault was in solidarity with a group (whose membership included medical practitioners) supporting abortion, the Groupe de l'information sur la santé.

5 The question can be added to Ewa Płonowska Ziarek's work on the reduction of women's bodies to bare life. In addition to regretting the omission in Agamben's work of considerations of slavery and of the rape of women, suggesting more generally that Agamben omits a consideration of both race and gender implications in his concept of bare life, Ziarek has proposed an analysis of the early-twentieth-century British radical suffragette movement and particularly its hunger strike strategies, whose stakes can be understood as a reappropriation by feminists of the woman's body reduced to bare life. See Ewa Płonowska Ziarek, "Bare Life on Strike: Notes on the Biopolitics of Race and Gender," *SAQ* 107:1 (2008): 89–105.

6 Foucault states: "What might be called a society's 'threshold of biological modernity' is situated at the point where the species has become the stakes of its own political strategies. For millennia, man remained what he was for Aristotle: a living animal with the additional capacity for a political existence; modern man is an animal whose politics places his existence as a living being [*sa vie d'être vivant*] in question." Michel Foucault, *The History of Sexuality*, vol. 1, *An Introduction*, trans. Robert Hurley (New York: Random House, 1978), 143. Translation modified. "Ce qu'on pourrait appeler le 'seuil de modernité biologique' d'une société se situe au moment où l'espèce entre comme enjeu dans ses propres stratégies politiques. L'homme, pendant des millénaires, est resté ce qu'il était pour Aristote: un animal vivant et de plus capable d'une existence politique; l'homme moderne est un animal dans la politique duquel sa vie d'être vivant est en question." Michel Foucault, *La volonté de savoir* (Paris: Gallimard, 1976), 188.

7 Giorgio Agamben, *Remnants of Auschwitz: The Witness and the Archive*, trans. Daniel Heller-Roazen (New York: Zone Books, 1999), 135.

8 Andrew Norris is discussing the neomort: "Here the reassertion of rights is simply not an option. We must decide whether a neomort—a body whose only signs of life are that it is 'warm, pulsating and urinating'—is in fact a human being at all, an agent or a thing." Norris, "Introduction: Giorgio Agamben and the Politics of the Living Dead," in *Politics, Metaphysics, and Death: Essays on Giorgio Agamben's "Homo Sacer"*, ed. Norris (Durham, NC: Duke University Press, 2005), 1–30, 14.

9 Thomas Carl Wall, "Au Hasard," in *Politics, Metaphysics, and Death*, 31.

10 Giorgio Agamben, *Homo Sacer: Sovereign Power and Bare Life*, trans. Daniel Heller-Roazen (Stanford, CA: Stanford University Press, 1998), 131.

11 Article 2 of the Enabling Act (*Ermächtigungsgesetz*) states: "Laws enacted by the gov-

ernment of the Reich may deviate from the constitution as long as they do not affect the institutions of the Reichstag and the Reichsrat. The rights of the President remain." Gesetz zur Behebung der Not von Volk und Reich (Law to Remedy the Distress of the People and the Nation), Reichstag, March 23, 1933, www.bpb.de/publikationen/ 04962540304433072098131403597315,9,0,Beginn_der_nationalsozialistischen_ Herrschaft.html#art9 (accessed June 26, 2007).

12   Giorgio Agamben, *State of Exception*, trans. Kevin Attell (Chicago: University of Chicago Press, 2005), 12.

13   See Judith Butler, "Indefinite Detention," in *Precarious Life: The Powers of Mourning and Violence* (London: Verso, 2004), 51. Butler writes: "In the name of a security alert and national emergency, the law is effectively suspended in both its national and international forms. And with the suspension of law comes a new exercise of state sovereignty, one that not only takes place outside the law, but through an elaboration of administrative bureaucracies in which officials now not only decide who will be tried, and who will be detained, but also have ultimate say over whether someone may be detained indefinitely or not."

14   Agamben, *State of Exception*, 3.

15   Uniting and Strengthening America by Providing Appropriate Tools Required to Intercept and Obstruct Terrorism Act of 2001 (USA Patriot Act), 107th Congress, H.R. 3162, public law 107-56, October 24, 2001. See http://thomas.loc.gov/cgi-bin/bdquery/z?d107: HR03162:@@@L&summ2=m& (accessed June 26, 2007).

16   However, Agamben notes that the Patriot Act did, prior to the November 13, 2001, military order, specify that within seven days detained aliens "had to be either released or charged with the violation of immigration laws or some other criminal offense." Agamben, *State of Exception*, 3. Also, there have been challenges to the act's application. For example, the targeting of homeless people in Summit, New Jersey, under the Patriot Act clause mentioning "attacks and other violence against mass transportation systems" was challenged, and there has been room for contestation of the act's extended application by individual plaintiffs and the American Civil Liberties Union.

17   *Roe et al. v. Wade, District Attorney of Dallas County*, 410 U.S. 113 (1972), October 11, 1972. See www.conlaw.org/cites2.htm (accessed June 27, 2007).

18   Mary Poovey, "The Abortion Question and the Death of Man," in *Feminists Theorize the Political*, ed. Judith Butler and Joan Scott (New York: Routledge, 1992), 244.

19   See www.conlaw.org/cites2.htm.

20   In Britain, abortion was a crime from 1803 onward, still illegal under the 1861 Offences against the Person Act. Through the twentieth century, increasingly broad exceptions were granted by the following: the Infant Life (Preservation) Act of 1929, which allowed term-limited abortions to protect the woman's life only; the Bourne Ruling of 1938, which extended the exception to include psychological grounds; and the Abortion Act of 1967, which consolidated the legality if there was a threat to the physical or mental health of the mother or existing children and if certified by two doctors. The Australian law was first governed by the British 1861 act. Despite its widespread availability (under an assortment of exceptions, including economic, social, and medical grounds, and usually with time limits), abortion has not been fully legalized in any state except the Australian Capital Territory, which passed the Abolition of Offence of Abortion Act in 2002. See 1861

Offences against the Person Act, OAP, 24 & 25 Vict., c. 94; Infant Life (Preservation) Act of 1929, 1929, 19 & 20 Geo., c. 34 5; *Rex v. Bourne* [1939] 1 K.B. 687, 3 All E.R. 615 (1938); Abortion Act of 1967, 15 & 16 Eliz. 2, c. 87 27, October 1967; and Crimes (Abolition of Offence of Abortion) Act 2002 (Act 2002 no. 24), September 9, 2002, www.legislation.act.gov.au/a/2002-24/20020909-2735/pdf/2002-24.pdf (accessed June 26, 2007).

21 Michèle Le Doeuff, *Hipparchia's Choice: An Essay Concerning Women, Philosophy, Etc.*, trans. Trista Selous (Oxford: Blackwell, 1991), 247. Translation modified. Le Doeuff stresses that women's control of their own fertility is not enshrined as a fundamental right. Though there have been changes to French abortion law since the publication of *Hipparchia's Choice*, the conditional nature of its legality has persisted.

22 Agamben, *State of Exception*, 7.

23 Ibid., 87.

24 See www.conlaw.org/cites2.htm.

25 Giorgio Agamben, *Means without End: Notes on Politics*, trans. Vincenzo Binetti and Cesare Casarino (Minneapolis: University of Minnesota Press, 2000), 6–7.

Andrew Benjamin

Particularity and Exceptions: On Jews and Animals

The animal is retained within both the history of philosophy and the history of art.[1] However, the nature of these relations and thus the conception of animality take on importantly different forms. Hence, relationality and animality have a history that is neither continuous nor organized within a perpetual sameness. While the animal has symbolic and representational presence, it is also the case that the animality in question has differing modalities. In a painting by Piero della Francesca of the archangel Michael having just slain the devil (*Saint Michael*, circa 1469), the saint is presented as having decapitated an animal. While the animal is, of course, the appearance of the devil, it is nonetheless unmistakably animal. In the biblical source, the devil oscillates between "dragon" and "snake." In this painting, however, the devil has nothing other than a snakelike quality. Having slain it, Saint Michael stands with the animal's head in one hand, and in the other, he holds his falchion. Neither the animal's face nor its body has traces or indications of being human. The reference, therefore, is to an intrinsic animality. The apparent nonchalance of Saint Michael's stance reinforces the position in which what obtains is not indifference but the enactment of a specific economy in respect to the animal. The dead animal operates in a domain in which its retention is structured by that economy. The human as the aftereffect of the "word" having become "flesh" reinforces, in this presentation, the incorporated refusal of the animal. As such it is one of a number of forms of animal presence.

The "same" biblical narrative occurs in Bartolomé Bermejo's painting *Saint Michael Triumphant over the Devil* (circa 1468). Nonetheless, in this instance the mode of presence is significantly different. Animality has a more complex register. While the devil in this work is a conglomeration of animals whose coordinated presence comprises its actual body, the body in question has a clear relation to the human. The reference, therefore, is no longer to an intrinsic animality. The proportion of the body, and this includes even the exaggerated mouth, is human. The second face beneath the dominant one has the structure of a human torso. The first of these faces has a mouth that, despite its size, has the same relation to nose, eyes, and ears as would be found on a human face. In the source text, namely, Revelation 12:7–12, the animal is named twice. It is both "dragon" and "snake." The event—Saint Michael fighting the "devil"—prompted the artwork. The provocation draws on the relationship between the words *dragon* and *snake*. While the terms are synonymous on one level, the snake denotes a form of malign cleverness that is not present in the dragon. The dragon, on the other hand, may allow human qualities to have visual presence. While Piero gives greater emphasis to the reference to the presence of evil in Genesis as opposed to the two images demanded by Revelation, the move from one iconic source to the other—the two paintings, in this instance—has a radically different registration in relation to the history of the animal.[2]

The works by Piero and Bermejo warrant detailed investigation in their own right; nonetheless, what they establish is a genuine difference between images of animal presence. In regard to the first, its particularity needs a setting. In this instance the animal's death saves humanity from the presence of evil. Human good, thus construed, takes as its ground the animal's death. This is, of course, no mere death. It is part of an economy that establishes human good. Moreover, once it is possible to argue that humanity comes to be what it is insofar as the human approximates to the image of God, then on the level of the image, what counts as being human incorporating the good proper to human being will itself be given within and thus secured by the operation of this economy. Within such an economy, human potentiality necessitates the death of the animal.

Particularity and Exceptions 73

Piero della Francesca,
Saint Michael (Panel from a Polyptych), circa 1469,
photo © The National Gallery, London

Bermejo's presentation of the devil opens in a different direction. Here, the animal and human combine in the creation of the devil. Hence, the animal has another significant presence within the history of art.[3] A confluence of the human and the animal in the presentation of the devil opens a more complex form of presence within the image, one that distances the straightforward conception of the economy demanding the animal's death. Bermejo's painting is not isolated. Albrecht Dürer's celebrated engraving *Knight, Death, and the Devil* (1513) presents the latter as the intersection of the human and the animal.[4] As with Bermejo, Dürer is able to acknowledge an already present possibility, namely, human animality. Indeed, it can be argued that Dürer's devil is even more human than Bermejo's. As a possibility, the animal is there initially to be overcome. And yet, its already being there—the original being there of the animal allowing at the same time the inevitable inscription of human animality—opens another possibility. In Dürer's engraving, the knight moves past both death and the devil. The sense of direction, a directionality evoking the copresence of the moral and the epistemological, gives centrality to the interplay of being human and a unidirectional path to be followed. Once that path is followed, the devil as the intersection of the human and the animal can then be excluded. That intersection is present as a "truth" about human being and equally as a warning. The truth is the insistent possibility that animality may take over. The warning is that the threat of the animal is counterposed to that which is proper to human being—being human, therefore, having a founding propriety. As a threat it demands the animal's continual excision. The warning, therefore, does not exist as a simple singularity if that means it need be given only once. While Saint Michael needed to kill the animal in order to secure that which is proper to human being, Bermejo's painting and Dürer's engraving reinforce the necessity for a form of continuity. Indeed, what they suggest is the need for vigilance against the threat of the animal. However, once continually present, that threat could always become a form of accommodation. In other words, what these two works stage is the possibility that the human and the animal—thus, human and nonhuman animals—cannot be simply divided, as though the excision and thus the difference had been decisively established. Rather than indifference, there is an always already present relatedness. What both works demonstrate is that within the human—indeed, constitutive of its very specificity—is a recalcitrant animality. At the beginning, therefore, there is not just another potentiality; rather, there is a significantly different sense of animal presence.

Albrecht Dürer, *Knight, Death, and the Devil*, 1513, image courtesy of British Museum/HIP/Art Resource, NY

What these artworks bring to the fore is a complex of concerns. In the first instance, it is the possibility that the animal is positioned as the other whose death reinforces and sustains human being. The economy sustaining this death (and its related conception of the animal) is as much bound up with the necessity for that death as with maintaining the animal's alterity. While there is one organizational logic at work within Piero's painting, Bermejo's painting and Dürer's engraving suggest another. In the case of the latter two, the animal cannot be given as simply the other to the human. Within this frame of reference, integral to the human is its presence as animal. Animality is part of being human. It is the nature of that presence and thus its relation to the definitions of human being that are central. The argument, therefore, is that being human is already to be with animals. Animality thus construed precludes the designation of neutrality. The question is how should we understand this state of what can be called "existing with animals (animality)"?[5]

First, there needs to be the recognition not just of an already present engagement with the animal but also that the engagement is articulated in terms of the complex of concerns opened by these artworks. What this complex includes, as noted, are two original and importantly different determinations. They should not be reduced to each other. Moreover, they already configure two of the dominant forms taken by the relationship between the human and the nonhuman animal. In the first instance, this particular configuration involves an economy in which the animal's differentiation from the human, let alone human animality, is inextricably bound up with the necessity of the animal's death. The death may be literal, for example, the dead snake in Saint Michael's hand. Equally, it could be a complete differentiation in which the animal is dead to "us." That death may be the animal's silence—silence in the realm of *logos*—though it could be the incorporation of the animal.

In the second, there is the transcription of the animal's original presence in a way that obviates the possibility of an equation of the animal with the necessity of its death. As such the economy of death that figured in the first instance would have become inoperable. There is, therefore, a division at the origin. The animal is already more than one; it is originally divided between these two possibilities. In addition, it is possible to argue that despite these clear divisions, each one recalls the other. As such there will always have been more than one animal; the *animal*—allowing the term an almost pragmatically abstract quality—is the more-than-one. Allowing for this setup will provide the way into Giorgio Agamben's engagement with the question of the animal in *The Open: Man and Animal*.[6]

Central to Agamben's analysis is the identification of what he describes as two "anthropological machines." What is significant about this description is that instead of simply positing relations between "man" and "animal," Agamben is concerned to note the way that relation is produced historically. (The history in question is as much concerned with philosophy and theology as it is with art and literature.) These machines stage the relationship between human and animal. Moreover, a different mode of production operates in the "modern" period than operated at an earlier stage. Regarding the modern version, Agamben formulates its presence thus: "It functions by excluding as not (yet) human an already human being from itself, that is by animalizing the human, by isolating the nonhuman in the human" (*O*, 37). This argument reappears, for Agamben, in relation to the Jew. Anti-Semitism draws upon the anthropological machine repositioning the Jew in terms of what is described by Agamben as "the non-man produced within the man" (*O*, 37). The earlier version of this machine—the machine producing the human/animal relation—operates in a "symmetrical" way. Within it, he argues, "the inside is obtained through the inclusion of an outside, and the non-man is produced by the humanization of the animal" (*O*, 37). For Agamben this latter position encompasses both the *homo ferus* and the slave. Within the formulation of Agamben's argument, the slave is the human form taken by the animal. Prior to moving to the next stage of the argument, it needs to be noted that this earlier version of the anthropological machine, one that would have produced Dürer's devil, is presented as bound up with the nonhuman. While that result will always be a possibility—that is, the production of the nonhuman—what Dürer's engraving suggests is that this produced entity cannot be separated in any absolute sense from the insistent presence of human animality. What emerges as a question, therefore, is how the ineliminable trace of that animality is to be positioned even if a version of Agamben's anthropological machine were to be accepted. In other words, to what extent could the production of the nonhuman in the human not have been marked in advance by the process that produced it? That mark—what would count as an original inscription—would allow for another sense of opening insofar as it would refuse the structure in which the separation of the nonhuman within the human is effected.⁷ From the beginning, equally at the beginning, there would be a mark. Its presence would undo, as a possibility, the divide, and thus the separation, within the human. It will be essential to return to this point. The decisive part of Agamben's argument is the move that he makes next.

The significant claim is that what allows both of these machines to operate is that they construct a "zone of indifference," which takes on the form of a caesura. The character of this zone, even its presence, is, however, the point to be contested. Agamben describes it as a "space of exception." He goes on to argue: "Like every space of exception, this zone is, in truth, perfectly empty, and the truly human that should occur there is only the place of a ceaselessly updated decision in which the caesura and their rearticulation are always dislocated and displaced anew. What would thus be obtained, however, is neither an animal life or a human life, but only a life that is separated and excluded from itself—only a bare life" (*O*, 38). "Bare life" is, of course, one of the dominant themes within Agamben's philosophical project.⁸

The strength of Agamben's argument lies in the provocative supposition that what allows for the operation of this anthropological machine is the construction at its interior of a zone of indistinction—in other words, a moment in which the division between animal and human is suspended, though a zone whose locus of operation is the machine itself. In *Homo Sacer* this position is presented once again in terms of the caesura. The point of absolute indecision is the camp.⁹ For Agamben, the camp, what amounts to the nomos of the modern, is itself defined as the place of exception. As such it is the place in which "the state of exception has become the rule."¹⁰

If there is a problematic element in Agamben's argument, then it concerns the positing of a zone of indetermination not just as a precondition for becoming determinant but, as significantly, as that zone having to be absolutely indeterminant.¹¹ Indeed, what I will suggest is that the contrary is the case. Rather than a caesura in which value is withdrawn, there is a porous site in which the relationship between self and other, the human and its posited other, and an alterity in which the animal would be inscribed are present as a continual site of negotiation. Allowing for the presence of that site opens up the animal to include within it human animality. It may be, therefore, that Dürer's engraving is closer to the truth than had been thought hitherto. What needs to be added is that any form of negotiation, even in relation to the deprivation of identity, occurs as a result of the complex determinations of power. The operation of power leaves its mark. This is true without exception. What is at issue, therefore, is the effect of this mark's retention. It is as though implicit within Agamben's overall argument is a form of utopianism in which harbored within the structure of the *homo sacer* is a neutrality that would configure the human beyond the hold of identity. It would be a utopianism premised on the erasure of this founding mark. The necessity of this mark—though equally the necessity of its erasure, as noted—works to establish limits.

It is in relation to the figure of the Jew that a fundamental aspect of the more general argument concerning the exception begins to emerge. As such it is essential to look in detail at one specific, and lengthy, formulation of this position in *Homo Sacer*. Within it Agamben argues the following:

> The wish to lend a sacrificial aura to the extermination of the Jews by means of the term "Holocaust" was . . . an irresponsible historiographical blindness. The Jew living under Nazism is the privileged negative referent of the new biopolitical sovereignty and is, as such, a flagrant case of a *homo sacer* in the sense of a life that may be killed but not sacrificed. His killing therefore constitutes . . . neither capital punishment nor a sacrifice, but simply the actualization of a mere "capacity to be killed" inherent in the condition of the Jew as such. The truth . . . is that the Jews were exterminated not in a mad and giant holocaust but exactly as Hitler had announced, "as lice," which is to say, as bare life. The dimension in which the extermination took place is neither religion nor law, but biopolitics.[12]

Fundamental to the formulation of this position is the identification of the Jew with bare life, that is, as "life separated and excluded from itself."

It is essential to be precise here. Bare life is the state of exception, and thus it is neither animal nor human. In the strictest sense, all determinations are withdrawn, and what emerges is a state to be determined. Hence, bare life discloses a space in which what awaits is the actualization of a potentiality, what is described in the text as "mere 'capacity to be killed.'"[13] That capacity—as a potentiality—inheres in life without determination, that is, in bare life. What cannot be resisted, therefore, is the question, who then are killed? The answer cannot be that it is simply the Jew in virtue of the Jew's capacity to be killed. That would be true of all humans and, indeed, of biological life in general. The answer needs to incorporate particularity. To put the position more emphatically, could there be an answer to the question that did not incorporate the founding mark? If the answer were to be in the affirmative, then it would sanction the possibility of a founding sense of particularity.

In this instance, the reason for pursuing the question of particularity can be located in what is noted above concerning the anthropological machine, the machine operated by "animalizing the human," which for Agamben amounts to the same thing as "isolating the non-human in the human" (O, 37). The animal, in terms of the possibility already at work in Dürer's engraving—namely, the presence of the animal within the human—unfolds in this direction. What needs to be examined is not the consistency of Agamben's argumentation but the possibility of either a state that is anterior to the human/animal or one structured by an indifference at the interior. In other words, what needs to be questioned is the possibility of this conception of the exception. Inherent within it is a conception of particularity without identity.

More recently, Agamben has returned to the structure of the state of exception. In this context it comes to be described as "a space devoid of law, a zone of anomie in which all legal determinations—and above all the very distinction between the public and the private—are deactivated."[14] This state of affairs is produced. As with bare life, as formulated in the same work, it is "a product of the biopolitical machine and not something that pre-exists it."[15] Rather, the interplay of the political and the body allows for bodies to occupy spaces created by the law's suspension. What produces bare life? This question is inextricably bound up with one posed earlier: who is killed? Once the question can be answered beyond simple generality, such that it will have become necessary to distinguish between potential and actual victims, then the identification of the production of bare life provides, at the same time, a ground of possible resistance that is other than universalism. Universality cannot account, philosophically or politically, for the difference between potential and actual victims. Highlighting causality may lead to a better understanding of the state of affairs described by Agamben, but it may equally, as indicated, begin to call into question the possibility of bare life as the site of absolute indistinction.

At the outset there can be no argument against a description of what occurs at Guantánamo Bay or even in Auschwitz in terms such that the inhabitants occupy spaces defined by the suspension of law. What matters with the suspension of law is not that it involves legislation that might be contestable. Rather, the interplay of the political and the body allows for bodies to occupy spaces created by the law's suspension. In other words, intrinsic to this setup is its possibility. Responding both to that possibility and to its actualization is not to respond in the name of law (where, of course, law is equated with statute). Such a conception of law will have been suspended. The reality of Auschwitz, though it should be conceded from the start that the simple evocation of this name is far from unproblematic, lies in the capacity for decisions linked to the elimination of certain groups.[16] Elimination necessitates a form of suspension. While the enacting that characterizes this procedure may involve the equation of Jews with lice, an equation in which the human comes to be reduced to the "non-human in the human," there is an actual sense of the specific at work. Movement has particularity—Jew to louse. (Movement is, of course, another staging of the general question of the relationship between the mark and singularity.) At Guantánamo Bay, the suspension of law equally involves movement. The identification of a range of individuals takes place such that the act of identification allows for the suspension and thus the creation of the exception. In both instances, there is an allowing. How, on a philosophical level, is this allowing to be understood? The question has an urgency precisely because the defense of law and humanity has already been countered by the reality of Guantánamo Bay, not because it is against the law to have acted in that way. But acting in that way involved the law's suspension, and hence, it is necessary to establish a ground of contestation, a moment in which the philosophical takes up the political as its direct concern.

Accounting for what is allowed returns us to the question of causality. Whatever quality bare life may have, it is produced. While the exercise of genuine political power (i.e., sovereignty) can be identified with the capacity to effect the movement that is the production of bare life, the movement in question is of necessity selective. To the extent that an explanation of the production of bare life cannot be given beyond a general claim concerning the anthropological machine, then what is removed is the possibility of accounting for particularity. Particularity will have been effaced by the machine's operation. Once such an account has to be given, then rather than the suspension of the law and the creation of the zone of complete indetermination, what appears more likely is that the movement of production—the causality proper to bare life—marks the presence of a matrix of concerns in which determinations always occur. The reason for holding to this description is that what has to endure is the necessity of being able to argue that what takes place—the reduction to bare life—occurs, for example, in relation to Jews or in relation to an already determined "enemy" (such as so-called Islamic militants). Those identified, the victims who become bare life, are positioned in advance. Bareness, therefore, is always a determination as an aftereffect. It operates by producing those who have already been identified as being subject to that process (i.e., to the process of subjectification). This determination means that sovereignty necessitates the capacity to discriminate. Discrimination occurs within a complex field of identities, which are attributed and constructed, on the one hand, and, on the other, may be regional and linked to versions of autonomy and affirmation. Sovereignty's capacity to position within such a complex is the definition of sovereign power, while at the same time it indicates that "bareness" is never completely bare. Discrimination leaves its mark.

There is an original determination precisely because there is a need to link individuation and discrimination. The mark of the Jew and the accusation of being a "terrorist" trace the bodies that were thought to have been neutral and thus may become "bare." This mark produces the distinction between the marked and the unmarked.[17] This distinction is fundamental if a conception of the "enemy" is to be maintained and, moreover, if such a conception is to have mobility. In this context mobility means that there will be different and thus new "enemies." Not only, therefore, does this mark differentiate, given that it is produced by the law's suspension, but it also accounts for why it is only in terms of particularity that the law can be suspended. The "state of emergency" does not simply occur but is inextricably bound up with the particularities that it produces. Identity is only ever the result of a complex process of individuation. As such those implicated within a situation in which the law is suspended are always marked by the deprivation of the suspended law. Once, through a process of production, the possibility of being a subject of right no longer pertains, then accounting for a process of subjectification in which subject and right are separated will ground resistance. For resistance to be effective, what needs to be understood is why that deprivation or separation has occurred, and part of that account demands paying attention to the specific.

At work here is a conception of identity in which there is a process of individuation. Only within such a complex can identities be continually produced. The interrelation between identity and production means that subjects are always the aftereffect of the system that produces them. For example, the production of the Jew as the "the non-man produced within the man" both individuates and differentiates. In other words, the Jew figures within such a production, while the other to the Jew (hence, the reciprocal production of Jew as this other) is differentiated from this Jew. It is, of course, this Jew who is killed rather than bare life. This Jew has died. Indeed, any further positioning—for example, the one that is called "bare life"—has to presuppose this initial movement. The additional element that has to be noted is that the production of this Jew, as with any further positioning based on it, is the effacing of a conception of difference (and related cultural practices) that has to assume its (difference's) ineliminability—a setup in which the ineliminable other is never absolute but always specific. What is assumed in such a position is a primordial relatedness. Therefore, once it is essential to hold to this sense of relatedness—a relation of porosity and negotiation defining self/other and human/animal relations—then Agamben's ontology, which refuses precisely this conception of founding differences, would, as a consequence, need to cede its place to a differential or relational ontology.[18] The positioning of the Jew as "the non-man produced within the man" has to be understood, therefore, as the refusal of exactly this latter conception of the ontological.

The production of identity entails that particularity is never an isolated occurrence. The excluded bear the mark not just of exclusion—a mark that could be no mark at all—but also the link between their particularity and exclusion. Assuming a primordial relatedness does not involve a return to the form of argumentation dominated by a concern with rights, as though rights functioned as ends in themselves. On the contrary, it assumes that within any relation lines of division are only ever porous and that relation necessitates that presence and modes of being present are always to be negotiated. To insist on porosity and negotiation is, therefore, the countermove. Porosity indicates that what can never be at work is the centrality of the human or even the definition of animality that takes the already positioned human as the point of departure.

If there is a politics implicit in Agamben's project, then it can be located in one of the final summations he provides in *The Open*. What for him becomes the response necessary to the operation of what has been called the "anthropological machine"—remembering that it is this machine that produces the animal as well as the Jew—entails rendering "inoperative the machine that governs our conception of man" (O, 92). He continues by arguing that this "will therefore mean no longer to seek new—more effective and more authentic—articulations, but rather to show the central emptiness, the hiatus that—within man—separates man and animal, and to risk ourselves in this emptiness: the suspension of the suspension, Shabbat of both animal and man" (O, 92). The significance of this passage will emerge from its juxtaposition with one of Agamben's earliest formulations of singularity without identity, the singularity that will become bare life, on the one hand, and Antonio Negri's recent discussion of Agamben's work, on the other.[19]

The "emptiness" alluded to is captured in the possibility of the community of what Agamben identifies as "singularities." While Agamben's description is of a state of affairs not tolerated by the state, it is this site of intolerance that defines the possibility of a community to come. At work here is the utopian impulse in Agamben's thought. The position is formulated in the following terms: "What the State cannot tolerate in any way is that the singularities form a community without affirming an identity, that humans co-belong without any representable condition of belonging (even in the form of a simple presupposition)."[20] The significance of this conception of community is clear. The position of the *homo sacer* will have been redeemed. It is this aspect of Agamben's work that Negri identifies when he argues that Agamben "ethically and conceptually goes beyond the state of exception by going through it: just as primitive christianity and the communisms of the origins had gone through power and exploitation and destroyed them by emptying them . . . Agamben's analysis shows how immanence can be realist and revolutionary."[21] *Immanence* is another way of describing the utopian impulse. For Negri this is the position that is opened up by Agamben's use of what Negri refers to as an "undifferentiated ontology." This ontological configuration characterizes the state of exception. Within it, to deploy Negri's formulation, "each element is reassumed in the empty game of an equal negativity."[22] The accuracy of this description is not at issue. What it brings into play is a further elaboration of the "emptiness." In Agamben's formulation, what is at risk is a version of "ourselves." And yet, what needs to be questioned is the "our" of "ourselves." Counterposed to a formulation of subjectification in terms of a community without identity, a "sacred" community, there is the recovery of a positioning in which this "our," thus "ourselves," will have always been more-than-one. This is a site of an original relatedness. The origin in question is an ontological position and not a locus of ethical obligation. This relatedness will be a relation to self as much as to the other and, therefore, equally to the other in ourselves.

Relation, therefore, brings back into play what was identified at an earlier stage as the already present more-than-one. On one level, this is the truth that was always there in Dürer's engraving, namely, that what can never be separated is the human and the animal—an impossibility that opens up the already present relation of self/other and human/animal. They would be fixed relations and thus constrained to be thought of philosophically in terms of the static rather than the dynamic were it not for porous borders yielding sites of negotiation. These sites and the complex of borders that are brought into play are the loci—places within becoming—that comprise the histories of alterity as well as the complex continuity of the animal's ineliminable presence. Allowing for both relatedness and porousness would mean that all that could ever be at risk within such an allowing is the residual humanism that posits, in this instance, the suspension of human animality rather than its continual affirmation.

Human animality has its most insistent presence in what Freud referred to as the "drives." At the center there are porous lines marking an impossible unity. This impossibility refuses melancholia since the only element to have been lost would be a retrospective projection of either a founding unity or a produced neutrality. Both have to be worked through. Rather than the language of emptiness, there needs to be the continual recognition of an ongoing incompletion. Activity and thus forms of practice take this founding sense of the incomplete as the point of departure. The porous is from the start that which cannot be completed; doing so would stem the movement it maintains. Negotiation as the site of decisions and responsibility has to be maintained as a site of continuous activity and, therefore, of cultural practices. The extension becomes clear. All lines that divide involve a form of separation that can be made absolute only after the event. This is not to posit a type of equality or even a sameness; it is rather to allow for continuities and differences. The question of the animal, allowing the continuity of movement between animal and animality, repeats the question of the other to the extent that what must be maintained are already present senses of relatedness. Particularity emerges only within those relations, within their retention and affirmation, and not with their suspension.

Notes

1. One of the most important and sustained accounts of the relationship between philosophy and the animal is that of Élisabeth de Fontenay, *Le silence des bêtes: La philosophie à l'épreuve de l'animalité* (*The Silence of the Beasts: Philosophy and the Challenge of Animality*) (Paris: Fayard, 1998).
2. The reference is to Genesis 3:1–13. It should be noted that in this context the "snake" speaks and is thus unlike any other animal. Moreover, the snake is cast out because of his actions. In other words, if the casting out created the distinction between God and Satan, then it is an occurrence that takes place as a result of both the human and the animal (though in this instance it is specifically the snake) sharing the very capacity that the philosophical tradition takes as dividing them, namely, language.
3. There are, of course, other possibilities. What could be contrasted with this depiction of the animal is the dog in Pietro di Cosimo's *A Satyr Mourning over a Nymph* (ca. 1495). Suspending symbolic registration for a moment, what appears in this work is the dog as observer. Other animals occur in the background. Presented either as detached observers or simply occupying the same space, animals have neither a negative nor a positive presence within a logic of sacrifice. The question of their relation endures as posed.
4. There is an important secondary literature on this engraving, but for the most part, it concentrates on the horse and the knight. Even Erwin Panofsky only notes in passing the "personification" of death and the devil. Regarding the latter, see Panofsky, *The Life and Art of Albrecht Dürer* (Princeton, NJ: Princeton University Press, 1995), 151–4.
5. This paper is extracted from a seminar, Existing with Animals, I have held at Monash University over the past two years. The project can be summed up as the attempt to resist the following position announced by Martin Heidegger in *The Fundamental Concepts of Metaphysics: World, Finitude, Solitude*, trans. William McNeil and Nicholas Walker (Bloomington: Indiana University Press, 1995). Heidegger argues in relation to being with animals that "this 'being-with' [*Mitsein*] is not an 'existing-with' [*Mitexistieren*] because a dog does not exist but merely lives [*ein Hund nicht existiert, sondern nur lebt*]" (308).

    Hence, the project of the seminar is to work through the consequences for philosophy were dogs to "exist" and thus a primordial existing with animals (animality) were to be allowed. Moreover, what happens to "life" when it is no longer opposed to existence?
6. Giorgio Agamben, *The Open: Man and Animal*, trans. Kevin Attell (Stanford, CA: Stanford University Press, 2004). Hereafter cited parenthetically by page number as O.
7. Central to the argument developed here is the connection between the mark and an original sense of relatedness. Clearly this formulation draws on the work of Jacques Derrida. In regard to the question of the animal, see *L'animal que donc je suis* (Paris: Galilée, 2006), 83. More generally in Derrida's work, see "Le retrait de la métaphore," in *Psyché: Inventions de l'autre* (Paris: Galilée, 1987), 63–95. The *trace*, the *mark*, and the *trait* are terms central to Derrida's mode of philosophical argumentation. The indebtedness here has its own limit. In this argument, the mark and a primordial relatedness are part of the terminology of a differential or relational ontology. Hence, the project has another direction.
8. The most sustained treatment of bare life is Agamben's work *Homo Sacer: Sovereign Power and Bare Life*, trans. Daniel Heller-Roazen (Stanford, CA: Stanford University

Press, 1998). I have offered a critical engagement with this concept in my "Spacing as the Shared: Heraclitus, Pindar, Agamben," in *Politics, Metaphysics, and Death: Essays on "Homo Sacer"*, ed. Andrew Norris (Durham, NC: Duke University Press, 2005), 145–72.
9 This position is worked out in a number of places in Agamben's writings. See, in particular, *Means without End: Notes on Politics*, trans. Vincenzo Binetti and Cesare Casarino (Minneapolis: University of Minnesota Press, 2000), 36–44.
10 Agamben, *Homo Sacer*, 169.
11 It may be that Agamben has addressed this point in *Homo Sacer* in relation to his discussion of Alain Badiou. In regard to that work, there is the suggestion that there is a relation that persists within both the process of exclusion and the creation of the exception (25). However, if this is the case, then it is incompatible with the later claim that it is a space "devoid of law." More significantly, it would assume a primordial relatedness and thus a potential undecidability within the decision that would undermine his arguments concerning indetermination.
12 Agamben, *Homo Sacer*, 114.
13 Ibid.
14 Giorgio Agamben, *State of Exception*, trans. Kevin Attell (Chicago: University of Chicago Press, 2005), 50.
15 Ibid., 87–88.
16 In this regard, see the exchange between Derrida and Jean-François Lyotard after the latter gave his paper "Discussions, ou: Phraser 'après Auschwitz'" at the Colloque de Cerisy in 1980. The proceedings of the colloquium, containing Lyotard's paper, were published as *Les fins de l'homme: A partir du travail de Jacques Derrida* (*The Ends of Man: From the Work of Jacques Derrida*), ed. Philippe Lacoue-Labarthe and Jean-Luc Nancy (Paris: Édition Galilée, 1981). The exchange occurs on pages 311–13.
17 The current practice of profiling at airports can be accounted for in these terms. In addition, it opens up the basis of understanding the significance of both disguise and produced identities. Regarding the latter, the essential literary work is Arthur Miller, *Focus* (London: Methuen, 2002).
18 For the conception of a differential ontology that informs this engagement with Agamben, see my *The Plural Event* (London: Routledge, 1994).
19 Antonio Negri, "The Ripe Fruit of Redemption," www.generation-online.org/t/negriagamben.htm (accessed March 14, 2007).
20 Giorgio Agamben, *The Coming Community*, trans. Michael Hardt (Minneapolis: Minnesota University Press, 1993), 85.
21 Negri, "Ripe Fruit of Redemption."
22 Ibid.

Ewa Płonowska Ziarek

Bare Life on Strike: Notes on the
Biopolitics of Race and Gender

One of the most important contributions of Giorgio Agamben's work to contemporary political philosophy is his concept of "bare life," which allows us not only to revise the Foucauldian theory of biopower but also to rethink the political contradictions of modernity. Despite its importance, Agamben's theory of bare life does not, however, sufficiently address two crucial questions: the problem of resistance and the negative differentiation of bare life with respect to racial and gender differences. It is these questions, I argue, that are at the center of any critical feminist engagement with his work. Thanks to Agamben's revision of biopolitics, it becomes clear that resistance cannot be limited to the contestation of the law or of power structures; in fact, one of the most pressing political questions raised by *Homo Sacer* is whether bare life itself can be mobilized by emancipatory movements.[1] The second issue we need to reconsider is the way bare life is implicated in the gendered, sexist, colonial, and racist configurations of the political and, because of this implication, how it suffers different forms of violence.[2] The central paradox bare life presents for political analysis is not only the erasure of political distinctions but also the negative differentiation, or privation, such erasure produces with respect to differences that used to characterize a *form of life* that was destroyed. In order to develop the possibilities of resistance and the negative determinations of bare life, I will supplement Agamben's genealogies of bare life with two political cases—the first one represented by Orlando Patterson's discussions of premodern and modern forms of slavery,[3] and the second one by the hunger strikes of militant British suffragettes at the beginning of the twentieth century.

To develop the paradoxes of bare life, let us begin with Agamben's definition of this concept. Reworking Aristotle's[4] and Hannah Arendt's[5] distinctions between biological existence (*zoē*) and the political life of speech and action (*bios*), between mere life and a good life, Agamben introduces in *Homo Sacer* his own interpretation and his own necessarily selective genealogy of bare life from antiquity to modernity. Stripped of political significance and exposed to murderous violence, bare life is both the counterpart to and the target of sovereign violence. To avoid misunderstanding, I would like to stress the point that is made sometimes only implicitly in Agamben's work and not always sufficiently stressed by his commentators: namely, the fact that bare life—wounded, expendable, and endangered—is not the same as biological *zoē*, but rather it is the remainder of the destroyed political *bios*. As Agamben puts it in his critique of Thomas Hobbes's state of nature, mere life "is not simply natural reproductive life, the *zoē* of the Greeks, nor *bios*" but rather "a zone of indistinction and continuous transition between man and beast" (*HS*, 109). More emphatically, in the conclusion of *Homo Sacer*, Agamben stresses that "every attempt to rethink the political space of the West must begin with the clear awareness that we no longer know anything of the classical distinction between *zoē* and *bios*" (*HS*, 187). To evoke Theodor Adorno, we could say that bare life, not only the referent but also the effect of sovereign violence, is damaged life, stripped of its political significance, of its specific form of life.

For Agamben, bare life constitutes the original but "concealed nucleus" of Western biopolitics insofar as its exclusion founds the political realm. Bare life is captured by the political in a double way: first, in the form of the exclusion from the *polis*—it is included in the political in the form of exclusion—and, second, in the form of the unlimited exposure to violation, which does not count as a crime. Thus, the most fundamental categories of Western politics are not the social contract or the friend and the enemy, but bare life and sovereign power (*HS*, 7–8). As Agamben's broad outline of political genealogy suggests, the position and political function of bare life change historically. This genealogy begins with the most distant memory and the first figuration of bare life expressed in ancient Roman law by the obscure notion of *homo sacer*—that is, the notion of the banned man, who can be killed with impunity by all but is unworthy of either juridical punishment or religious sacrifice. Neither the condemned criminal nor the sacrificial scapegoat and thus outside both human and divine law, *homo sacer* is the target of sovereign violence exceeding the force of law and yet anticipated and authorized by that law. Banished from collectivity, he is the referent of the sovereign decision on the state of exception, which both confirms and suspends the normal operation of the law. In Agamben's genealogy, the major shift in the politicization of bare life occurs in modernity. With the mutation of sovereignty into biopower, bare life ceases to be the excluded outside of the political but in fact becomes its inner hidden norm: bare life "gradually begins to coincide with the political realm" (HS, 130). However, this inclusion and distribution of bare life *within* the political does not mean its integration with political existence; rather, it is a disjunctive inclusion of the inassimilable remnant, which still remains the target of sovereign violence. As Agamben argues, "Western politics has not succeeded in constructing the link between *zoē* and *bios*" (HS, 11).

In contrast to the ancient ban, or the inclusive *exclusion* from the political, the new form of disjunctive *inclusion* of bare life within the *polis* emerges with modern democracies. In democratic regimes, this hidden incorporation of bare life into both the political realm and the structure of citizenship manifests itself, according to Agamben, as the inscription of "birth" within human rights—an inscription that establishes dangerous links among citizenship, nation, and biological kinship. As the 1789 Declaration of the Rights of Man and of the Citizen proclaims, men do not become equal by virtue of their political association but are "born and remain" equal. Democratic citizens are thus bearers of both bare life and human rights; at the same time, they are the targets of disciplinary power and free democratic subjects. In a political revision of Michel Foucault's formulation of modern subjectivity as an "empirico-transcendental" doublet,[6] Agamben argues that the modern citizen is "*a two-faced being, the bearer both of subjection to sovereign power and of individual liberties*" (HS, 125). The democratic subject of rights is thus characterized by the aporia between political freedom and the subjection of mere life, without a clear distinction, mediation, or reconciliation between them.

Since bare life is included within Western democracies as their hidden inner ground (*HS*, 9) and as such cannot mark their borders, modern politics is about the search for the new racialized and gendered targets of exclusion, for the new living dead (*HS*, 130). In our own times, such targets multiply with astonishing speed and infiltrate bodies down to the cellular level: from refugees, illegal immigrants, inmates on death row subject to suicide watch, comatose patients on life support, to organ transplants and fetal stem cells. For Agamben, this inclusion of bare life within each citizen's body becomes catastrophically apparent with the reversal of democratic states into totalitarian regimes at the beginning of the twentieth century. As the disasters of fascism and Soviet totalitarianism demonstrate and as the continuous histories of genocide reveal, by suspending a human, political form of life, totalitarian regimes can reduce whole populations to disposable bare life that can be destroyed with impunity. According to Agamben, this is what constitutes the unprecedented horror of the Nazi concentration camps: the extreme destitution and degradation of human life to bare life subject to mass extermination. "Insofar as its inhabitants were stripped of every political status and wholly reduced to bare life, the camp was also the most absolute biopolitical space ever to have been realized, in which power confronts nothing but pure life, without any mediation" (*HS*, 171). If Agamben controversially claims that the concentration camp is not just the extreme aberration of modernity but its "fundamental biopolitical paradigm" (*HS*, 181), which shows the "thanatopolitical face" of power (*HS*, 142, 150), it is because concentration camps for the first time actualize the danger implicit in Western politics, namely, total genocide made possible by the reversal of the exception signified by *homo sacer* into a new thanatopolitical norm. Such a collapse of the distinction between exception and norm, together with the "absolute" and unmediated subjection of life to death, constitutes the "supreme" political principle of genocide (*HS*, 142).

The most compelling force of Agamben's work is his diagnosis of the ways the aporia of bare life and form of life in Western politics gives rise to new forms of domination and to the catastrophic turns of history, which culminates in the thanatopolitics of fascism. Nonetheless, Agamben's analysis of this aporia from antiquity to modernity misses two crucial issues: the question of resistance and the negative differentiation of bare life along racial, ethnic, and gender lines. First of all, as argued by several commentators and critics, most notably Ernesto Laclau, what is lacking in Agamben's work is the theory of "emancipatory possibilities" of modernity.[7] Yet, if we were to reconstruct such a theory in terms of Agamben's philosophy, then the task of conceptualizing resistance could not be limited to the contestation of the law or power structures; in fact, one of the most important political questions is whether bare life itself can be mobilized by oppositional movements. By focusing on the way bare life functions as the referent of the sovereign decision, Agamben, unfortunately, answers this question in the negative: "The 'body' is always already a biopolitical body and bare life, and *nothing in it* . . . seems to allow us to find solid ground on which *to oppose the demands of sovereign power*" (*HS*, 187; emphasis added). The second problem Agamben ignores is the way bare life is implicated in the gendered, sexist, colonial, and racist configurations of biopolitics. If we argue that bare life emerges as the aftereffect of the destruction of the symbolic differences of gender, ethnicity, race, or class—differences that constitute political forms of life—this means that bare life is still negatively determined by the destruction of a *historically specific* way of life. Thus another paradox of bare life is a simultaneous erasure of the political distinctions and negative differentiation retrospectively produced by such erasure.

Let us consider these two issues—the differentiation of bare life and its role in emancipatory movements—in turn. Although Agamben's heterogeneous examples of bare life—for instance, the father-son relation in antiquity, Nazi euthanasia programs for the mentally ill, the destruction of the Romany, ethnic rape camps in the former Yugoslavia, Karen Quinlan's comatose body, and especially the most important case of the *Muselmann*—are always diversified along racial, gender, and ethnic and historical lines, his conceptual analysis does not follow the implications of such heterogeneity. Consider, for instance, his brief comment about the difference between ethnic rape camps and Nazi camps: "If the Nazis never thought of effecting the Final Solution by making Jewish women pregnant, it is because the principle of birth that assured the inscription of life in the order of the nation-state was still—if in a profoundly transformed sense—in operation. This principle has now entered into a process of decay" (*HS*, 176). Needless to say, the sexually and racially marked difference between these two forms of sovereign violence—genocide and rape—cannot be reduced to the principle of birth alone. Agamben refrains from any further explorations of rape as sexual political violence because such an analysis would complicate his very concept of bare life, always defined in relation to death and not to sexual violation.

To demonstrate the need to supplement Agamben's conceptualization of bare life, I would like to consider briefly two historical cases—the first represented by Aristotle's and Patterson's discussions of slavery, the second by the British militant suffragettes' writings on the hunger strike. In terms of Agamben's genealogy of bare life, slavery is an important case to consider because its racialized ancient and modern forms represent instances of bare life coextensive with both the Greek *polis* and modern democracy and yet irreducible to the category of either *homo sacer* or the camp. Let us begin the exploration of bare life and slavery with the text that is foundational to Agamben's political theory, Aristotle's *Politics*.[8] As soon as Aristotle introduces the crucial distinctions between *zoē* and *bios*, *oikos* (home) and *polis*, he is confronted with the localization and legitimation of enslaved life, which does not seem to fit easily into these classifications. Thus, not only is it the case that, as Thomas Carl Wall argues, in the Greek *polis* bare life "was abandoned to the home, the *oikos*,"[9] but a more fundamental problem is that Aristotle's defense of slavery creates a conceptual aporia that undermines his definition of slavery as an "animate instrument" belonging to the household.[10] Implicated in the network of differences fundamental to the differentiation of the public space of the city—such as the differences between the body and the soul, the male and the female, the human and the animal, passion and reason—enslaved life, defined by Aristotle as property, does not have a "proper" place. In his apologia, Aristotle writes: "The soul rules the body with the authority of a master: reason rules the appetite with the authority of a statesman. . . . The same principle is true of the relation of man to other animals. . . . Again, the relation of male to female is naturally that of the superior to the inferior. . . . We may thus conclude that all men who differ from others as much as the body differs from the soul, or an animal from a man, . . . are by nature slaves."[11] As these multiple analogies show, the political subjection and exclusion of femininity and slavery are "like" the subjection of the body to reason and animality to humanity. Perhaps bearing witness to the threat of enslavement in war, this analogy potentially makes the body of the Greek citizen "like" the enslaved or inhuman body. Conversely, the enslaved body blurs the distinction between the human and the animal, the household and the city. Because of its in-between position on the "threshold" (to deploy Agamben's frequently used term in *Homo Sacer*), slavery in Aristotle's text begins to haunt the Greek *polis* from within and from without, making the Greek citizen, prior to its modern counterpart, already *"a two-faced being, the bearer"* of enslavement to reason and a political being among equals (HS, 125).

Although subjected to the violence of the master rather than to sovereign banishment, enslaved life in Aristotle's *Politics*, like the obscure figure of *homo sacer* in Roman law, blurs the boundaries between the inside and outside of the political. It is Patterson's influential study of slavery from antiquity to modernity that gives a full account of the liminality of the slave's paradoxical position in the social order. In *Slavery and Social Death*, Patterson argues that the enigma of slavery exceeds both the juridical and the economic categories of law, production, exchange, or property. What all these categories fail to explicate is both "total" domination of the enslaved life and the liminality of the slaves' position. Like the indistinction, or the threshold, between inside and outside marked by *homo sacer*, a slave's liminality collapses both the political and the ontological differences between human and inhuman, monstrosity and normality, anomaly and norm, life and death, cosmos and chaos, being and "nonbeing" (SSD, 42, 44). In one of the most suggestive passages, which is devoted to the interpretation of the Anglo-Saxon representation of slavery and servitude in *Beowulf*, Patterson writes, "It was precisely because he was marginal, neither human nor inhuman, neither man nor beast, neither dead nor alive, the enemy within who was neither member nor true alien, that the slave could lead Beowulf and his men across the deadly margin that separated the social order above from the terror and chaos of the underground" (SSD, 48).

What is, then, the relation between these two expressions of subjugation and liminality, represented by *homo sacer*, on the one hand, and by enslaved life, on the other? The concept that links bare life and sovereignty with the master/slave dialectic is the substitutability of slavery for death: either for the death of the external enemy or the death of the internal "fallen" member of the community. According to Patterson, this substitution of enslavement for death is echoed in the "archetypal" meaning of slavery as social death (SSD, 26). Such a substitution does not give pardon but, on the contrary, creates the anomaly of the socially dead but biologically alive and economically exploited being. Because the expropriation of a slave's life constitutes him or her as a nonperson or a socially dead person, it produces another instance of bare life, violently stripped of genealogy, cultural memory, social distinction, name, and native language, that is, of all the elements of Aristotle's *bios*. Akin to secular excommunication, slavery in all its different historical formations was institutionalized as the extreme destruction of the sociosymbolic formation of subjectivity. This extreme mode of deracination and exclusion from symbolization, the *polis*, and kinship reconstituted enslaved life as a nameless, invisible nonbeing, "as *pro nullo*" (*SSD*, 40).

The notion of slavery as a substitute for death complicates Agamben's central thesis that sovereign decision/bare life constitutes the foundational political paradigm in the West. First, although the extreme delegitimation and the nullity of enslaved life make it another instantiation of bare life, the very fact that such life undergoes substitutions of one form of destruction for another undermines from the start the centrality of *just one* paradigm of politics. Second, slavery raises the question of whether the destruction of the historically specific form of life is a "condition" of exchangeability as such. As Patterson argues, the destruction of the political forms of life turned human beings into "the ideal human tool . . . perfectly flexible, unattached, and deracinated" (*SSD*, 337). Because of its fungibility, such a "disposable," "ultimate human tool" (*SSD*, 7) is also a perfect commodity, and indeed, Patterson notes instances in which slavery functioned as money. We can argue, therefore, that the violent production of social death functions as a hidden territory not only of politics but also of commodity exchange. Consequently, the substitution of social death for biological death indicates a possible transformation of the sovereign ban into ownership and exchange. As Patterson's discussion of the ancient Roman doctrine of *dominium* suggests, absolute power merges with the absolute ownership of *res* (*SSD*, 30–32).

What both slavery and *homo sacer* have in common is the production of bare life stripped of its historically specific form of life, and yet what distinguishes them is the contrast between the sovereign ban and the marginal inclusion of enslaved life. If the sovereign decision on the state of exception captures bare life in order to exclude it, the biopolitics of slavery is confronted with the profitable inclusion of socially dead beings. Hence, Patterson argues that after the stage of violent depersonalization the next stage of enslavement introduces "the slave into the community of his master, but it involves the paradox of introducing him as a nonbeing" (*SSD*, 38). Since, unlike *homo sacer*, the socially dead being has to be included within and made profitable, this second stage of the biopolitics of slavery poses the dilemma of "liminal incorporation" (*SSD*, 45). The paradox of liminal incorporation is the opposite of the sovereign exclusion, even though it creates similar effects of indistinction. In place of a sovereign decision on the state of exception, we have institutionalized containment within the law of a permanent anomaly, which confounds the differences between life and death, destruction and profit.

In a reversal of the slaveholder's absolute domination into parasitical dependence, Patterson rewrites the Hegelian master/slave dialectic—which explains such dependence in terms of the desire for recognition—as "human parasitism" (SSD, 334–39). This reversal has another crucial consequence that is downplayed in Agamben's theory of sovereignty: parasitical dependence provides a new ground on which to theorize the possibility of resistance and emancipation. The emphasis on resistance, which negates a prior destruction of forms of life and calls for the creation of new forms, culminates in Patterson's claim that the most important political discovery of enslaved peoples is that of freedom: "The first men and women to struggle for freedom, the first to think of themselves as free . . . were freedmen. And without slavery there would have been no freedmen" (SSD, 342). Although Patterson is deeply troubled by making enslavement even a contingent condition of freedom, his insistence on the ongoing struggle for liberation by dominated people points to another legacy of modernity sidestepped by Agamben: the legacy of revolutionary and emancipatory movements.

Agamben is right that the praxis of liberation calls for the ontology of potentiality. Yet he never considers potentiality from the perspective of bare life—that is, from the perspective of the impossible—focusing instead on the often obliterated difference between potentiality and sovereign power. What makes it especially difficult for him to theorize emancipation in any greater detail are the parallels he establishes all too quickly between potentiality, event, the excess of the constituting power, and sovereign exception. In his polemic, Agamben claims that there are in fact no grounds to distinguish between revolutionary praxis and sovereign exception: "The problem of the difference between constituting power and sovereign power is, certainly, essential. Yet the fact that constituting power neither derives from the constituted order nor limits itself to instituting it—being, rather free praxis—still says nothing as to constituting power's alterity with respect to sovereign power" (HS, 43). Perhaps Agamben does not see any criterion by which to distinguish transformative praxis from sovereign violence because he is primarily concerned with the topological excess of sovereign violence vis-à-vis the political order. As he admits, "The question 'Where?' is the essential one once neither the constituting power nor the sovereign can be situated wholly inside or altogether outside the constituted order" (HS, 42). However, if we switch the terms of the analysis from "where" to "how"—that is, from Agamben's topology to the most important Foucauldian lesson about techniques of power—then the difference between transformative praxis and sovereign violence becomes more apparent. Although both types of power exceed the constituted order, their mode of operation is different. The excess of sovereign power manifests itself as a suspension of the law, as the exclusion of bare life, as a state of exception that either confirms the norm or, in extreme cases, collapses the distinction between the exception and the norm. The mode of operation of the transformative power, however, is not the decision on the exception but the negation of existing exclusions from the political followed by the unpredictable and open-ended process of creating new forms of collective life—a process that in certain respects more closely resembles an aesthetic experiment rather than an instrumental action.

As I have suggested, another reason Agamben does not consider the practice of liberation in any greater depth is that his ontology of potentiality is developed to undermine sovereign will and not to transform bare life—the configuration of the impossible—into a site of contestation and political possibility. To theorize the notion of bare life as a contested terrain, I would like to turn now to another political case—to the British suffragettes' use of the hunger strike at the beginning of the twentieth century. This case reveals once again three interrelated aspects of bare life: its negative differentiation with respect to the politics of race and gender; its subjection to different forms of violence; and its role in multiple emancipatory movements. Let me begin with the facts that tend to be all too easily taken for granted. At the turn of the twentieth century, racialized and gendered subjectivities still occupied liminal positions in Western democracies and as such were associated in the political imaginary with the inclusive exclusion of bare life. Yet these subjectivities were also the "bearers" and the creators of a very different legacy of modernity, that of multiple liberation movements. In this context, the suffragettes' hunger strikes can be regarded as an invention of a mode of political contestation, which mobilizes bare life for emancipatory struggle. Consequently, this case allows us to supplement Agamben's analysis in a crucial way: a hunger strike not only reveals the hidden aporia of democracy—the aporia between the politicization of bare life as the object of biopower and political freedom guaranteed by human rights—but it also shows how this aporia can enable revolutionary transformation.

Although the history of hunger strikes is often obscure, they were practiced in ancient Rome, medieval Ireland, and India as a means of protest, frequently to exert moral pressure or to force a debtor to return his debt.[12] After the Easter Rising of 1916 in Ireland, the hunger strike was adopted in the Irish struggle for independence in 1917,[13] and it was most famously employed by Mohandas Gandhi, who fasted at least fourteen times in British-occupied India.[14] Nonetheless, it was militant British suffragettes who in 1909 revived and redefined the hunger strike as a modern political weapon of an organized movement by linking it for the first time with the discourse of human rights. The political practice of hunger striking in suffrage agitation was initiated by suffrage militant, painter, and artist Marion Wallace Dunlop, who was arrested and sentenced to one month of imprisonment for having written on the wall of Parliament an extract from the English Bill of Rights.[15] Dunlop began a hunger strike to protest the denial of the status of the political offender, and after ninety-one hours of fasting, she was released because prison officials, ignorant about the effects of the hunger strike, were afraid she would become a martyr for suffragettes. By the time other suffragette prisoners were released before the expiration of their sentences, the hunger strike had been adopted by members of the suffrage movement as an effective political weapon both to terminate prison sentences and to create new possibilities of revolt within the disciplinary apparatus of the prison. In response to this unprecedented act of protest, after King Edward VII's personal intervention in August 1909, Home Secretary Herbert Gladstone ordered the striking suffragettes to be force-fed—a brutal punitive retaliation, which up to that point had been practiced primarily in insane asylums.[16]

How can we understand this configuration of the hunger strike as a weapon of resistance and the sadistic brutality of forcible feedings? Although this was one of the most dramatic episodes in the struggle for women's suffrage,[17] hunger strikes and the political reprisals of forcible feeding are still undertheorized means of democratic protest. In his study of nonviolent political action, Sharp classifies the hunger strike as a means of political intervention demanding a transformation of power relations and a redress for injustice.[18] For Kyra Landzelius, the hunger strike is a "corporeal challenge" to the "discursive practices of power."[19] As suggested by Lady Constance Lytton's letter to the *Times*, written on behalf of herself and eleven other hunger-striking suffragettes on October 10, 1909, the hunger strike is both a protest and a demand for new freedoms, an appeal articulated through the double, sharply disjoined medium of a publicly circulating letter and the starving body secluded in prison and barred from public appearance. In her letter, Lytton claims that subjugated groups resort to violence against their bodies when rational law-based arguments fail—that is, when instituted political speech is deprived of its performative power: "We want to make it known that we shall carry on our protest in our prison cells. We shall put before the Government by means of the hunger-strike four alternatives: to release us in a few days; to inflict violence on our bodies; to add death to the champions of our cause by leaving us to starve; or, and this is the best and only wise alternative, to give women the vote. We appeal to the Government to yield, not to the violence of our protest, but to the reasonableness of our demand."[20] Lytton's emphasis on the "violence" of the hunger strike seems paradoxical: such violence, inflicted on the self as a substitute target for political power, acts by refusing to act; it collapses clear distinctions between passivity and activity, actuality and potentiality, victim and enemy. On the one hand, the hunger strike repeats, mimics, and exposes in public the hidden irrational violence of the sovereign state against women's bodies. On the other hand, by usurping the state's power over bare life, the "nonact" of self-starvation negates women's exclusion and calls for the transformation of the law. By usurping sovereign power over bare life, hunger-striking women occupy both of these positions—the sovereign and *homo sacer*—at the same time, and this is what distinguishes their status from comatose patients, the inmates of concentration camps, that is, from all those beings that, in extreme destitution, are reduced to bare life alone. What is thus performed in the hunger strike is the collapse of the distinctions between sovereignty and bare life, will and passivity, potentiality and actuality, the struggle for freedom and the risk of self-annihilation. Maud Ellmann rightly calls such a performance a "gamble with mortality."[21] And as the word *gamble* implies, at stake here is a transformation of the central opposition between the sovereign decision and bare life into radical contingency in political life.

Although not analyzed by Agamben, the emphasis on the collective political struggle over bare life is an important element in Lytton's January 31, 1910, speech, delivered at the Queen's Hall, only a week after her release from prison. This address defines the hunger strike as a weapon against the political enemy:

> People say, what does this hunger-strike mean? They will not realize that we are like an army, that we are deputed to fight for a cause, . . . and in any struggle or any fight, weapons must be used. The weapons for which we ask are simple, a fair hearing, but that is refused us. . . . Then we must have other weapons. What do other people choose when they are driven to the last extremity? . . . They have recourse to violence. . . . These women have chosen the weapon of self-hurt to make their protest.[22]

In response to antisuffrage propaganda, Lytton argues that hunger strikes are not unreasonable attacks of hysteria but a calculated choice of the last resort by the "army" of the dispossessed. As acts of "warfare" by the paradoxical means of self-injury and refusal, hunger strikes allowed suffragettes to continue their revolutionary struggle without directly engaging in war. Furthermore, by extending the possibility of militancy from the public sphere to prison itself, the hunger strike changes imprisonment into a new means of "fighting for a cause," transforms punishment into rebellion, turns subjection into the ambiguous political agency of self-hurt.

The most suggestive way Lytton's speech evokes the notion of bare life as a new weapon of oppositional movements is through the figurative juxtaposition of feminine, animal, and divine bodies. Her speech begins with an analogy between a degraded female body, deprived of rights, and a deformed animal body, abused on its way to the slaughterhouse, and she ends by contrasting the tortured body of imprisoned suffragettes to Christ's sacrificed body. Unlike the sacrificial lamb with which Christ is frequently compared, the deformed sheep, a powerless "creature" mistreated by "the crowd," is the very opposite of either a human or a divine sacrifice.[23] Designating the passage between the animal and the human, the "old and misshapen"[24] sheep is the figure of damaged life, deprived of political or religious significance—a life whose biological survival is at risk. When in a sudden insight Lytton discovered this hidden analogy between femininity and the deformed animal life, she decided to join the militant suffrage movement—a decision that transformed her life and gave it political and collective meaning. We can plumb the depth of this transformation by contrasting the frightened isolated animal, powerless to protest its abuse, and the "army" of women forming a revolutionary movement in order to fight for access to the political.

The suffragettes' usurpation of the sovereign decision over mere life in the struggle for political rights negates their exclusion and suspends the current law, at least on the symbolic level. Yet this act does not constitute a state of exception, which, through the act of exclusion, establishes the normal frame of reference or, as in the case of fascism, turns exception into a new norm. Rather, suffrage militancy represents a revolutionary call for a law yet to come. As Landzelius argues, the hunger strike stages a political trial of the existing law and political authority. In this "meta-juridical trial," the private act of starvation reverses the guilty verdict imposed on the militant suffragettes into a public condemnation of the government.[25] Thus, the hunger strike perverts juridical punishment into a means of interrogating the law itself and contesting the government's authority. By reversing the roles of the defendants and the accusers, the hunger strike performs a double chiasmatic transfer between bare life and the law, between the present and the future. On the one hand, it transforms the private act of starvation into a collective contestation of the law; on the other hand, it summons the yet nonexistent authority of the new law by risking the physical life of the body. In a catachrestic movement, bare life anticipates what is unpredictable and beyond anticipation: a new law and a form of life of female bodies. In so doing, it transforms impossibility into potentiality.

As a counter to the sovereign decision, hunger-striking suffragettes seized hold of their bare life, wrested it away from sovereign decision, and transformed it into a site of the constitution of a new form of life. The suffragettes' public redefinition of the female body so that it no longer bore the repressed signification of bare life and acquired instead a political form not only challenged the sovereign decision over bare life, but in so doing called for a new mediation of life and form outside the parameters of that decision. At stake here is a new type of link between bare life and political form that would be generated from below, as it were, rather than imposed by a sovereign decision. As Wall argues, it is the absence of the relation between bare life and its politically qualified ways of life that calls for sovereign decision: "Bare life is nonrelational and thus invites decision. It is the very space of decision . . . and, as such, is perpetually *au hasard*."[26] By contesting a sovereign decision on bare life, the new link between bare life and forms of living cannot be confused with either a dialectical reconciliation or a celebration of prepolitical life. At the end of *Homo Sacer*, Agamben only hints at what this new form of mediation supplanting sovereign decision might look like: "This biopolitical body that is bare life must itself instead be transformed into the site for the constitution and installation of a form of life that is wholly exhausted in bare life and a *bios* that is only its own zoē. . . . We give the name form-of-life to this being that is only its own bare existence and to that life, that, being its own form, remains inseparable from it" (*HS*, 188). The key point here is inseparability and yet nonidentity between form and life, which make both their severing and their unification equally impossible.²⁷

As this discussion of the biopolitics of race and gender shows, such reconsideration of bare life in the context of racial and sexual politics calls for some fundamental revisions of that concept. As we have seen, bare life cannot be regarded in complete isolation from all cultural and political characteristics. If bare life emerges as the remnant of a destroyed form of life, then, according to Agamben's own emphasis on its *inclusive exclusion* in the political, its formulation has to refer, in a negative way, to the racial, sexual, ethnic, and class differences that used to characterize its *form* of life. In other words, bare life has to be defined as the remnant of a specific form of life that it is not yet or is no longer. Furthermore, bare life cannot always be considered as the exclusive referent of the sovereign decision, but it has to be reconceptualized as a more complex, contested terrain in which new forms of domination, dependence, and emancipatory struggles can emerge. By analyzing bare life as the target of sovereign violence, Agamben allows us to diagnose new forms of domination and political dangers in modernity. Although any praxis of freedom is dependent on such a diagnosis, at the same time such praxis exceeds the constituted forms of power and requires reflection on the often occluded role of bare life in another paradigm of democratic modernity—that of the struggle for freedom. In doing so, it transforms impossibility into contingency in political life.

Notes

1 Giorgio Agamben, *Homo Sacer: Sovereign Power and Bare Life*, trans. Daniel Heller-Roazen (Stanford, CA: Stanford University Press, 1998).
2 Catherine Mills is one of the very few Agamben interpreters to raise the question of sexuality and sexual embodiment but in the context of Agamben's theory of testimony rather than his theory of bare life. Mills, "Linguistic Survival and Ethicality," in *Politics, Metaphysics, and Death: Essays on Giorgio Agamben's "Homo Sacer"*, ed. Andrew Norris (Durham, NC: Duke University Press, 2005), 215–8. Even though Sidi Mohammed Barkat does not directly engage Agamben, his analysis of "le corps d'exception" of the colonized can be read as an illuminating "postcolonial" critique of Agamben's philosophy. Barkat, *Le corps d'exception: Les artifices du pouvoir colonial et la destruction de la vie* (Paris: Editions Amsterdam, 2005). Diane Enns's careful analysis of the ambiguities of revolt of the occupied bodies reduced to bare life is another crucial extension of Agamben's work

in the colonial context. Enns, "Bare Life and the Occupied Body," *Theory and Event* 7.3 (2004), http://muse.jhu.edu/journals/theory_and_event/toc/archive.html#7.3 (accessed November 23, 2006).

3   Orlando Patterson, *Slavery and Social Death: A Comparative Study* (Cambridge, MA: Harvard University Press, 1982). Hereafter cited parenthetically by page number as *SSD*.

4   In *Politics*, Aristotle makes a famous distinction between mere life and a good life to define the function of *polis*: "While it comes into existence for the sake of mere life, it exists for the sake of a good life." Aristotle, *Politics*, trans. Ernest Barker, ed. R. F. Stalley (Oxford: Oxford University Press, 1998), book 1, ch. 2, p. 10.

5   Hannah Arendt follows the Aristotelian distinction between *zoē* and *bios* in a number of her texts, most notably in *The Human Condition*, where she identifies the political life not only with speech and action but most importantly with the condition of human plurality. Arendt, *The Human Condition*, 2nd ed. (Chicago: University of Chicago Press, 1998), 7.

6   See Michel Foucault's discussion of "the empirical and the transcendental" in *The Order of Things: An Archaeology of the Human Sciences* (New York: Vintage, 1973), 318–21.

7   According to Ernesto Laclau, the absence of the theory of resistance is intertwined with the lack of the theory of hegemony. Laclau argues that Agamben fails to distinguish between totalitarian and democratic sovereignty, emerging from the hegemony of democratic movements. Laclau, "Bare Life or Social Indeterminacy?" in *Giorgio Agamben: Sovereignty and Life*, ed. Matthew Calarco and Steven DeCaroli (Stanford, CA: Stanford University Press, 2007), 11–22. For a different critique of the lack of attention to resistance in the context of the body and the contingency of political struggles, see also Andreas Kalyvas, "The Sovereign Weaver," in *Politics, Metaphysics, and Death*, 112–3.

8   I am grateful to my colleague Kalliopi Nikolopoulou for discussing with me Aristotle's notion of slavery. For her excellent discussion of the relation between Agamben and Plato, see "Between Art and *Polis*: Between Agamben and Plato" (unpublished, Buffalo, NY, 2006).

9   Thomas Carl Wall, "Au Hasard," in *Politics, Metaphysics, and Death*, 39.

10 Aristotle, *Politics*, book 1, ch. 4, p. 13.

11 Aristotle, *Politics*, book 1, ch. 5, p. 16.

12 For a brief discussion of the history of the hunger strike, see Gene Sharp, *The Politics of Nonviolent Action* (Boston: Porter Sargent, 1973), 363–67.

13 Maud Ellmann argues that the Irish nationalists might have been inspired by suffrage, but in order to conceal this, they appealed to the medieval practice of fasting against debtors to compel them to repay their debts. Ellmann, *The Hunger Artists: Starving, Writing, and Imprisonment* (Cambridge, MA: Harvard University Press, 1993), 11–12.

14 Sharp, *Politics of Nonviolent Action*, 637.

15 Midge Mackenzie, *Shoulder to Shoulder: A Documentary* (New York: Alfred A. Knopf, 1975), 110.

16 Lisa Tickner, *The Spectacle of Women: Imagery of the Suffrage Campaign, 1907–1914* (Chicago: University of Chicago Press, 1988), 104.

17 According to Jane Marcus, it is "perhaps the primary image in the public imagination regarding the 'meaning' of the suffrage movement." Marcus, "Introduction: Re-reading the Pankhursts and Women's Suffrage," in *Suffrage and the Pankhursts*, ed. Marcus (London: Routledge, 1987), 2.

18 Sharp, *Politics of Nonviolent Action*, 359.
19 Kyra Marie Landzelius, "Hunger Strikes: The Dramaturgy of Starvation Politics," in *Einstein Meets Magritte: Science, Nature, Human Action, and Society*, vol. 3, *Man and Nature: A World in Transition*, ed. Diederik Aerts (Dordrecht: Kluwer, 1999), 83.
20 Mackenzie, *Shoulder to Shoulder*, 135.
21 Ellmann, *The Hunger Artists*, 21.
22 Cheryl R. Jorgensen-Earp, *Speeches and Trials of the Militant Suffragettes* (London: Associated University Presses, 1999), 108–9.
23 Ibid., 108.
24 Ibid., 107.
25 Kyra Marie Landzelius, "Hunger Strikes," in *Encyclopedia of Food and Culture*, vol. 2, ed. Solomon H. Katz (New York: Charles Scribner's Sons, 2003), 220.
26 Wall, "Au Hasard," 39.
27 Giorgio Agamben, *Remnants of Auschwitz: The Witness and the Archive*, trans. Daniel Heller-Roazen (New York: Zone Books, 1999). In his discussion of the survivors' testimonies in *Remnants of Auschwitz*, Agamben defines such a link as the aporetic task of witnessing. For my discussion of the ethical structure of the survivors' testimony in *Remnants of Auschwitz*, see "Evil and Testimony: Ethics 'after' Postmodernism," *Hypatia* 18 (2003): 201–3.

Claire Colebrook

Agamben: Aesthetics, Potentiality, and Life

The philosophy of Giorgio Agamben is eclectic, both in terms of the writers on whom he draws—Martin Heidegger, Carl Schmitt, Michel Foucault, Walter Benjamin, Georges Bataille, Dante, Aristotle, Plato, Jacques Derrida, Ludwig Wittgenstein, and Immanuel Kant—and in terms of the topics he addresses, everything from the work of art to the constitution of the political. Despite this diversity of themes and philosophical traditions, one problem orients Agamben's various theoretical excursions: the problem of life and the loss or forgetting of that moment when life emerged in its human distinction. In philosophical terms, this is the problem of potentiality: how is "man" as a speaking, deciding being possible, and if "man" does *not* actualize his potential to speak—as occurs in Auschwitz, where he is reduced to the condition of mere being—how do we deal with this pure potential and its capacity *not* to become human? Agamben's project, then, is not a history of ideas but a determined retrieval, such that we might consider, once again, the coming into being of human life, its emergence into actuality from potentiality.

The central concern of this essay, however, is the relation between Agamben's theory of art and the concept of potentiality in its human manifestation. Potentiality forms the key concept of Agamben's specific mode of politics, for we should not simply accept what *is* but look at how the world where we live came into being. What renders the world of the human and art possible, and if such a world emerges from a potential, what other worlds and other modes of the human are possible? For Agamben it is the work of art that *should* disclose this pure potentiality. Today, however, our notion of art as nothing more than the object created by a will precludes us from recognizing art as disclosive of potentiality. Agamben's meditations on the history of aesthetics are most explicit in their mourning a loss of distinction: there was a time when humanity was defined by *poiesis*, a "doing" that brought the world into the truth of its presence. In modernity, however, all action is praxis: not the bringing forth of something other than the human but the pure act of the artist. The art object *is* "a Warhol" or "a Renaissance masterpiece." Art today either is mere potential for enjoyment or is valuable only insofar as it is the product of an irreducible will; there is no sense of the (once essentially human) power to produce art as other than *mere* life, as the opening of a world *other than the human as it already is, but nevertheless of human making*. Art functions for Agamben, then, as a site of loss (for art is now a mere product rather than a revelatory act) and as a site of redemption (for only art can reveal what politics has covered over). Today, when politics has become "means without ends," all we have is the world as so much potentially manipulable and manageable life; what we do not have is some idea or end as to what that potentiality might create. Even Agamben's seemingly metaphysical concerns, such as his writings on the different senses of potentiality in Aristotle, are motivated by a historical project of *retrieving* and *restoring* the emergence of a distinction or difference from life. This is why Agamben's work on aesthetics is always more than an examination of any single work of art but devolves on art's potential to revive an original openness to experience—a fuller life that is not yet the life *of humankind*.

From *Homo Sacer* to *The Open*, Agamben's work looks at a present distinction, such as the difference between the political and the nonpolitical or between the human and the animal, and retrieves the more original potentiality from which such differences are actualized. Despite this seemingly radical openness and despite his criticism of a modernity that is dominated by a concept of life as potential (in the sense of availability for action), I argue that Agamben's work is dominated by a normative image of life that is also a normative aesthetics. What he rejects is the increasing animalization of humanity as a mere willing that maintains itself without thought for the end it might bring into being. For Agamben, humankind is not differentiated from animals biologically; on the contrary, the key difference is between biological life, which lives only to maintain itself, and human poetic life, which lives in order to create forms. Humans, therefore, should be regarded as those beings who may or may not give themselves a world; they should certainly not be managed as so much (actual) bodily life that is a part of the world.[1] Concomitantly, a work of art is not an object produced by a subject, not a thing on which we may or may not gaze; it is certainly not a "meaning," which is then given some form.[2] Rather, before there can be an actual subject *who intends or perceives*, there must be the pure event of "saying," which may allow one to then say this or that (*RA*, 65). Art would then be the means by which the human opens to ends beyond the present, beyond what is given as one's own. Only if we see life *not* as an object that we may or may not view but as that which opens into subject and object, then will we also see art as the possibility of reopening the world. A *proper* conception of life as potentiality yields a proper conception of the work of art. Life, for Agamben, is considered in its radical potentiality, not when we simply accept the human as a being alongside the animal, but when we look at the emergence of the human from animality. That emergence, which occurs through poetic speech and artwork, is not something that occurs inevitably, as though potentials would always arrive at their fulfillment, for the human may not arrive.

We can see this nonarrival in Auschwitz, where the human becomes simply a body; in *homo sacer*, where the human is no more than that which must not be killed; and in the modern (fallen) artwork, where art is simply circulating production. On the one hand, then, Agamben is critical of the notion that humanity's potentiality must arrive at its fulfillment, and here he follows Heidegger in arguing that if we want to understand the human in its specificity we should look not for some achieved essence but at all those cases in which humans fail to arrive at themselves. If humans can become mere biological matter to be managed, what does this tell us of the fragility of human freedom? On the other hand, while being critical of any human essence or propriety, Agamben does insist on a proper mode for the work of art: art should not be a mere thing or object that is the chance outcome of an event but should be world disclosive. Whereas Agamben has challenged ontotheological humanism or the idea that the human is the animal that inevitably brings itself into being and arrives at its full potential, he shifts that proper potential to the work of art. Art should be the revelation of pure potentiality, not an object for consumption and certainly not a "product" detached from the revelation of the open.

The idea of a potentiality that properly gives itself life is tied to a traditional image of man and, particularly, man in the image of God. Agamben criticizes the fall into Christian theology and the conception of God as *actus purus*, an action that has no end outside itself. But his own appeal to a more proper life, a life that does not just circle around and maintain itself, a life that creates and brings forth what is not itself, resonates with a highly gendered (and theological) humanism. Indeed, one might argue that Agamben's sense of loss—that there was a time when life was truly creative, not yet the same dull round of production—must condemn as not properly alive that which merely happens or occurs. Current practices of installation art that rely on the contingencies of the environment would not, for Agamben, be disclosive of the world or of potentiality. Humanity is, and always has been, *properly* more than formless potential without end; man *becomes human* in creating and bringing forth an end that is not yet given. From Aristotle to the Renaissance, the masculine has been associated with form-giving power and, therefore, with what *truly is*, while the feminine as mere matter is what requires the life of spirit in order to be. What is repressed is what has always been figured as "woman": a birth, production, or creation that is neither expressive, free, nor open, a production all too weighed down by the already actualized. A creation that has no meaning or truth and that occurs without intent or appropriation: this is the monstrosity of real birth that has always had to be refigured, as Agamben will do, as properly an active bringing forth. So, while Agamben criticizes the modern notion of life as simply willing to maintain itself as it is, he draws on a supposedly lost higher sense of life as that which creates from itself in order to be more than itself: divine, poetic life—the life of man.

The Detachment of the Work of Art

According to Agamben, previously there had been an experience of art in which the producer was not in the position of one who must impose form on his material; instead, art unfolded from a common "dwelling" in the world: "In these epochs, the subjectivity of the artist was identified so immediately with his material—which constituted, not only for him but also for his fellow men, the innermost truth of consciousness—that it would have appeared inconceivable to speak about art as having value in itself" (*MC*, 33). For Agamben, the "aesthetic" in ancient thought referred to sensibility in general and only became restricted to the fashioned work of art in modernity (*MC*, 6). Agamben aims to retrieve this earlier breadth by destroying the "work" of art: "Perhaps nothing is more urgent—if we really want to engage the problem of art in our time—than a *destruction* of aesthetics that would, by clearing away what is usually taken for granted, allow us to bring into question the very meaning of aesthetics as the science of the work of art" (*MC*, 6). Agamben's appeal to both life and art, despite his professed defiance of a complacent humanism that accepts the human as already constituted, nevertheless privileges the artwork as an exemplary site of becoming, a becoming in which the human—so precariously demarcated from animal embodiment—once again encounters a becoming human. It is this coming to humanity that also aligns the work of art with speech, the praxis that constitutes the subject through its own action.

It is Agamben's commitment to a concept of potentiality that explains the crucial status of the work of art in his politics. According to Agamben, politics opens with a distinction between *bios* and *zoē*, between the human as a political being and mere biological life. But, for Agamben, it is just this distinction, which relies on the human as a theoretical being, that covers over the question of *how* the human comes to be defined as having a proper being. As modernity becomes increasingly biopolitical, "man" is no longer the being who decides his humanity; his humanity or *bios* as already actualized determines who may or may not engage in the political decision. What is lost in this increasing normalization of biopolitics is the life that is neither human nor inhuman: the open potentiality from which the speaking, self-constituting human will emerge. In the modern biopolitics of Auschwitz, an actualized norm of humanity—white, Aryan—decides in advance who will or will not count as human. For Agamben, one can overcome the logic of the camp not by expanding this logic—or including the Semitic peoples *as human*—but by recalling that the human does not have its being determined in advance. Indeed, it is in the camp that we see once again a life that has not yet actualized its humanity: "Auschwitz marks the end and the ruin of every ethics of dignity and conformity to a norm. The bare life to which human beings were reduced neither demands nor conforms to anything. It itself is the only norm; it is absolutely immanent" (*RA*, 69). Humanity is not reducible to any actual norm that might then govern political relations; rather, what we need to consider is the genesis of the normative, or the opening up of worlds, and it is here that art becomes important. Only a resurrected art, according to Agamben, will yield the properly nonrelational: art is neither the expression *of* an already existing subject nor the realization *of* an implicit humanity that must come to realization. Art is open relationality, a relation or potential *to*, where the infinitive is not governed by anything that already *is*. Art releases us from both a politics and an ontology of substance—*what is*—to a politics of potentiality: a future of open, unimpeded becoming.

But if art is the site of our redemption, we also need to approach the artwork with a sense of nonrelational potentiality. Only this will take us *back* from the modern artwork to the creativity of *poiesis*, or the event of the world's coming into being. What we need to do, Agamben insists, is to rethink the relation between potentiality and actuality and, also, to rethink the concept of the political beyond relationality.[3] So what is the nonrelational? Agamben gives an explicit account of the nonrelational in his essay on potentiality in Aristotle, and it is to this *proper* understanding of potentiality that we now turn.

According to Agamben, Aristotle rightly passes over a generic understanding of potentiality (or what is merely possible) in order to define a notion of potentiality proper—the latter being a specific and defining potentiality.[4] With proper potentiality, there is no remnant of potential once the passage to actuality has occurred: "What is truly potential is thus what has exhausted all its impotentiality in bringing it wholly into the act as such" (P, 183). For Agamben, who takes up Aristotle's notion of a proper potentiality that, when actualized, is fully actualized, it is speech that is humans' exemplary, specific, and defining potentiality. All humans have the potential to speak, and when they do, *this* potential is fully actualized. It is true that one can say this or that. One can then go on to say other things, even speak other languages, but all this would still be speech and not a different actualization of the potential. If one were *not* to speak by remaining silent, this would still be a positive mode of *having* the potential to speak. A potential that one *properly* has is a potential that may *not* be actualized, such that *not* speaking is still a possibility of this potential. (We do not say of a quiet animal that it is not speaking or being silent.)

If we were to say that a human is a political animal, a being who becomes what he or she is through speech and politics, how are we to understand this potentiality? The problem of bare life, according to Agamben, is that it has been understood as a potentiality in the generic or improper sense. Politics, in the Western biopolitical trajectory of sovereignty, has always been understood as the system of laws, relations, and actions *made possible by* a life that must properly realize itself in this external form, which inevitably passes from being mere life to political life. The relational—our political being and our linguistic being—is therefore grounded in our biological being, which is understood as before all other relations: as generic, animal, and unspecified potential, as the matter on which political form is grounded and realized (*HS*, 10). What Agamben's historical retrieval aims to achieve is the emergence of political being—of our realization of ourselves as beings of speech, art, and law—from animal being, and we must do this without seeing man as an animal that inevitably and essentially actualizes itself politically. Potentiality is properly, for Agamben, given in *poiesis* or the opening of a space enabling the human to view himself not as an actual being but as the possibility of viewing and form-giving as such: "Man has on earth a poetic status, because it is poiesis that founds for him the original space of his world" (*MC*, 101).

For Agamben, then, a good polity would not be organized according to the proper end of man. The proper polity is just this confrontation with potentiality—that we are not a potential *for* this or that end but create our ends from ourselves, with our "self" having no essence or proper potential other than this capacity to create without end. One is most human, or fully actualizing one's potential, when one is nothing other than one's potentiality. A polity liberated from a (biological) potentiality, which would merely be the ground or material for the formation of the polity, yields "extraterritoriality," freed from nativity or nation.[5] This is a polity in which the political is not reified into what we must become; becoming as such is constantly happening:

> Only if I am not always already and solely enacted, but rather delivered to a possibility and a power, only if living and intending and apprehending themselves are at stake each time in what I live and intend and apprehend—only if, in other words, there is thought—only then can a form of life become, in its own factness and thingness, *form-of-life*, in which it is never possible to isolate something like naked life.[6]

We can make more sense of this pure, nonrelational potentiality through Agamben's understanding of both speech and art. In *Homo Sacer* Agamben establishes an analogy between the passage to sovereignty and the passage to speech. Just as it is in the act of excluding mere life that the polity establishes itself as the body *other than life*, as *zoē*, so in speaking one establishes the outside of language *as* the prelinguistic: "Language is the sovereign who, in a permanent state of exception, declares that there is nothing outside language and that language is always beyond itself" (*HS*, 21). Any specific speech act presupposes an instituted set of relations, a *langue*. What we forget, however, is that both *langue* and *parole* presuppose a potentiality to speak. This potentiality needs to be grasped as pure, because nonrelational, existence: "Only language as the pure potentiality to signify, withdrawing itself from every concrete instance of speech, divides the linguistic from the nonlinguistic and allows for the opening of areas of meaningful speech" (*HS*, 21). That is, through speaking, in *this* act or *parole*, one must assume a lawfulness or *langue* that determines this speech as speech, as that which must have sense or be interpreted as bearing a meaning beyond itself. In each speech act, there is, therefore, a generality or translatability *and* a pure event of speaking (before one determines *what* I mean or say, there is the event *that* I am saying), the pure potential to speak. Insofar as I speak, I am already submitted to a system, but the system is just the necessary assumption that what I say has some sense beyond itself: "The non-linguistic, in this sense, is nothing other than a presupposition of language" (*P*, 71). Any word in a language is always a predication *of* this or that determined thing, but the event of speaking as such—that there is speaking—is simply itself: "It is precisely by means of the anonymity and insubstantiality of linguistic Being that philosophy was able to conceive of something like pure existence, that is, a singularity without real properties" (*P*, 71).

This is the only pure existence we can grasp. Every form of being is given *as* this or that entity, determined by some meaning or other. We *can* regard speech as the expression of some content, but we can also consider the being of the act of speaking, and it is here that we are given a coming into existence that is not the existence of any determined content. For Agamben, pure nonrelational existence is given in *that I am speaking*: this speaking that actualizes me as one who speaks does not leave anything outside itself (*P*, 21). We understand speech properly, then, not as the expression of who I am, behind the act, but as the act that produces me as one who speaks, brings me into existence *as such*. Here, Agamben is at his most theological and existential. (For, as Sartre notes, the desire to be pure being-in-itself and being-for-itself is the desire to be God.[7]) If Sartre's criticism of the inauthenticity of reducing human becoming to some form or essence paved the way for a political aesthetics of engagement, then Agamben's insistence on speech as pure existence, as open potentiality, recalls a history of theological political aesthetics that aligns the divine with a creation that is completely open and *unnecessary*—free *not* to become. Whereas all other essences must come into being or actuality in order to be realized, there is nothing that determines God *as pure existence* to realize himself; it is always possible that he not be. As Etienne Gilson notes, Thomas Aquinas distinguished himself from essential ontotheologies—defining God as exemplary of a certain concept or predicate, God as goodness—by inaugurating an existential ontology: God is a pure act of being, through which any essence, form, or potentiality is actualized.[8]

For Agamben, however, it is speaking man who realizes this ideal: we may express this or that concept *in speech*, but speech as such is the coming into existence of essence. If the God who is *nothing* outside his act of existence came to be an image of the modernist artist, as one who is nothing other than creation and not the creation *of* this or that,[9] Agamben's political human also bears this aesthetic normativity, to be nothing other than a creativity opened only from itself. What Agamben adds to this tradition of the political human as self-creating existence is the insight that one can become aware of this potential to exist, create, and spring forth from oneself only when this potential is *not* realized. For this reason, in addition to the work of art—exemplifying humans as creative, existing, and space inaugurating—one also needs to confront humans in their nonexistence.

Accordingly, the "remnant" of Auschwitz—the silent *Muselmann*—is neither one who is simply deprived of humanity nor one who is fully human and possessed of human rights. He evidences humanity in his *not* being allowed to become human. The *Muselmann's not speaking* is precisely what speech must bear witness to. What is witnessed is pure nonexistence, not the failure to say this or that but the absence of saying as such. All witnessing leaves behind this remaining potentiality: the potentiality not to be. The potentiality is there in its nonbeing. Linguistics, as a science, can understand this or that system of relations, this or that realization of our potential, but only philosophy, Agamben insists, can understand linguistic potentiality as such (P, 76). This potentiality comes before the potential/actuality or the bare life/political life oppositions; for it is only with this potential to speak, to relate to oneself from oneself, that anything like actuality or the political can be said. Against deconstruction, then, which has consigned the world to text, force, or relations (HS, 54), Agamben insists that we need to intuit *and experience* the potential from which relationality emerges: "The task that our time imposes on thinking cannot simply consist in recognizing the extreme and insuperable form of law as being in force without significance. . . . Only where the experience of abandonment is freed from every idea of law and destiny . . . is abandonment truly experienced as such" (*HS*, 60–61). It is this experience of the emergence of the relational that Agamben celebrates in the work of art. Thus, Agamben concludes *The Open* with a discussion of a Titian canvas (*The Three Ages of Man*) in which two lovers encounter each other with a gaze that is not that of subject to subject but of body to body that ever so tentatively opens the relation that allows subjectivity to come into being.[10] This work of art, then, has an exemplary value and concludes Agamben's meditations on a history of Western culture that has all too readily accepted the human as a being and not as one who must bring himself into being (and must do so properly and openly in the work of art). The turn to Titian at the conclusion of *The Open* is significant for two reasons. First, for all his talk about art, Agamben finds the proper image of potentiality *in* the artwork. The two lovers who gaze at each other are at once desired bodies, while the loving gaze also sees the other's enigmatic and not fully grasped being as a subjectivity that remains outstanding. It is the work of art, here, that *represents* humanity not as situated within relations but as opening up relations. Second, in addition to discussing a work of art that depicts the human, why does Agamben turn *back* to art before Marcel Duchamp? He does so because modern art has lost *poiesis*, or a genuine bringing forth. This is where we can see art as relying on a proper concept of world disclosure. Art is not of this world for Agamben; it is the giving of the world that returns us to potentiality.

Humans are, in essence, poetical—the point from which both facticity (or bare life) and the political (relational, historical, and ethical life) spring forth: "That art is architectonic means, etymologically: art, *poiesis*, is production (τίκτω) of origin (ἀρχή), art is the gift of the original space of man, *architectonics* par excellence" (*MC*, 100). In our increasingly biopolitical age, what we have lost is this life of potentiality, so dominated have we become by a life that is nothing more than what has already been actualized. Whereas for Heidegger it is being-toward death that discloses nonrelational potentiality[11]—and this because death takes everyday life beyond its already given possibilities to an end that is not given—for Agamben nonrelational potentiality is intimated as the other of a fallen politics, a fallen art, and a fallen space:

> Artistic subjectivity without content is now the pure force of negation that everywhere and at all times affirms only itself as absolute freedom that mirrors itself in pure self-consciousness. And, just as every content goes under it, so the concrete space of the work disappears in it, the space in which once man's action and the world both found their reality in the image of the divine, and in which man's dwelling on earth used to take its diametrical measurement. (*MC*, 56–7)

There is, then, for Agamben, a need to destroy the aesthetic, to get away from an art that is seen as the product of some action or the work of an artist (*MC*, 47) and instead think of art *poetically* as the disclosure of presence, the opening of space in general. What cannot be admitted, in Agamben's good polity and new space of humanity, is the fragmentation of potentiality: the idea that human life might be composed of, and traversed by, a series of powers not its own—unassimilable, opaque modes of production that are without world or synthesis.

For Agamben there *is* a proper life, a life of speech and self-departure that would yield, properly, both a proper aesthetics and a proper political space. Such an aesthetic would not be practical, the formation of a determined self, but poetic, the disclosure of a world (*MC*, 71). And this world could not be this or that territory, the possibility of this or that body or bodies, but possibility as such—a space of open and therefore nonexclusive potentiality. Far from perceiving a world of nations across which biopolitics might range because all nations are manifestations of the one human life, Agamben looks forward to a space freed from territoriality, an open, self-defining, and poetic space—a space of humanity's own making. Space, like art and the polity, must properly spring forth from potentiality in general and not just be the space across which this or that actual body is placed. Agamben, therefore, refuses an aesthetics of *praxis*, which would merely negotiate a field of forces where one is already within produced relations, and insists on an aesthetics of *poiesis*, where space and relations are produced from the one nonrelational potentiality, the human potential from which all other powers are disclosed:

> The central experience of poiesis, pro-duction into presence, is replaced by the question of the "how," that is, of the process through which the object has been produced. In terms of the work of art, this means that the emphasis shifts away from what the Greeks considered the essence of the work—the fact that in it something passed from nonbeing into

being, thus opening the space of truth (ἀ-λήθεια) and building a world for man's dwelling on earth—and to the *operari* of the artist, that is, to the creative genius and the particular characteristics of the artistic process in which it finds expression. (MC, 70)

For Agamben the political is properly poetic, and the poetic is properly political. Potentiality should depart from itself *not to create itself but to disclose the world, the space of man*. A human should never be produced as a producer but always remain just that which gives itself through the act of production, never a determined term in a relation as much as the potential for relations. Agamben's *poetic* space is, as he insists, extraterritorial, a "whatever" polis,[12] and it can only be grasped poetically as that which precedes action and constituted space. It is not surprising, therefore, that Agamben's exemplary consideration of a work of art—Titian's *The Three Ages of Man*—both restores and overturns an ideal of the aesthetic at the heart of Western politics. Since romanticism, the appeal to artists as legislators is an appeal to the opening of politics; what we take to be given, immutable, unchanging, natural, essential, or human is potentially open to re-creation if only we can step back from an already constituted humanity. Art is, for Agamben, neither a representation of the properly human as political, nor a pure freedom from all propriety, but an opening of the space where the two enter into relation. The Titian canvas figures *and* brings to presence this play between body and relation. On the one hand, the lovers have—in their bodily love—abandoned their mystery, allowed the other to experience their inhumanity. On the other hand, this is the profound mystery, the mystery of that which is *not* taken up by humanity in general:

> To be sure, in their fulfillment the lovers learn something of each other that they should not have known—they have lost their mystery—and yet have not become any less impenetrable. But in this mutual disenchantment from their secret, they enter . . . a new and more blessed life, one that is neither animal nor human. . . . In their fulfillment, the lovers who have lost their mystery contemplate a human nature rendered perfectly inoperative—the inactivity [*inoperosità*] and *desœuvrement* of the human and of the animal as the supreme and unsavable figure of life.[13]

What I would set against Agamben's appeal to the exemplary work of art is the thought of the inhuman, not as the potential at the heart of the human, but as that which has always been outside the aesthetic. If we consider a different object of art, one that is not a work or putting into relation, we might open the space not of man but of woman. For has woman not always been conceived as that matter or being that must be brought to existence and spirit through pure form? As an "example" I could cite the recent "act" of the Dundee, Scotland, art group Artillery, whose members on April 24, 2004, painted letters in water along a walkway of Dundee streets, letters that almost immediately disappeared and left no trace, except for the mention of the work here (and other possible but not inevitable memories). Or we could consider the sound installations of the Australian composer David Chesworth, whose audiovisual "Proximities" on Melbourne's William Barak Bridge occasionally "includes" the sound of summer insects, depending on the weather.[14] Here, art is subjected to the world, does not "open" us to another world, but presents us with a world whose givenness cannot be returned to a pure and creative potentiality that might be retrieved by poetic man. Such an artwork allows us to consider an idea of the aesthetic, not as the space that brings the potentiality to be to presence, not as reflective representation or production of a scene, but as event that does not preserve its own existence.

Notes

1. Giorgio Agamben, *Remnants of Auschwitz: The Witness and the Archive*, trans. Daniel Heller-Roazen (New York: Zone Books, 1999), 64. Hereafter cited parenthetically by page number as *RA*.
2. Giorgio Agamben, *The Man without Content*, trans. Georgia Albert (Stanford, CA: Stanford University Press, 1999), 12. Hereafter cited parenthetically by page number as *MC*.
3. Giorgio Agamben, *Homo Sacer: Sovereign Power and Bare Life*, trans. Daniel Heller-Roazen (Stanford, CA: Stanford University Press, 1998), 47. Hereafter cited parenthetically by page number as *HS*.
4. Giorgio Agamben, *Potentialities: Collected Essays in Philosophy*, ed. and trans. Daniel Heller-Roazen (Stanford, CA: Stanford University Press, 1999), 179. Hereafter cited parenthetically by page number as *P*.
5. Giorgio Agamben, *Means without End: Notes on Politics*, trans. Vincenzo Binetti and Cesare Casarino (Minneapolis: University of Minnesota Press, 2000), 25.
6. Ibid., 9.
7. Jean-Paul Sartre, *Being and Nothingness: An Essay on Phenomenological Ontology*, trans. Hazel E. Barnes (London: Methuen, 1956).
8. Etienne Gilson, *The Christian Philosophy of St. Thomas Aquinas* (London: V. Gollancz, 1957).
9. Claire Colebrook, *Gender* (London: Palgrave, 2003).
10. Giorgio Agamben, *The Open: Man and Animal*, trans. Kevin Attell (Stanford, CA: Stanford University Press, 2004), 86.

11 Martin Heidegger, *Being and Time*, trans. John Macquarrie and Edward Robinson (New York: Harper and Row, 1962), 295, 251.
12 Giorgio Agamben, *The Coming Community*, trans. Michael Hardt (Minneapolis: University of Minnesota Press, 1993).
13 Agamben, *The Open*, 87.
14 *Proximities* is a soundscape built up from David Chesworth's and Sonia Leber's recordings of the singing voices of people from fifty-three Commonwealth nations now living in Australia. The work is located on the William Barak Bridge in Melbourne. See www.waxsm.com.au/proximities.htm (accessed June 22, 2007).

Lee Spinks

Except for Law: Raymond Chandler, James Ellroy, and the Politics of Exception

Politics, Law, and the State of Exception

In his book *Homo Sacer: Sovereign Power and Bare Life*, Giorgio Agamben offers a radical reformulation of the modern relationship between sovereign power, politics, and "bare life," or what he calls "the simple fact of being."[1] The nexus of power, politics, and life inevitably recalls Foucault's influential discussion of a politics of life (or *biopolitics*) in which the disciplinary power of the modern state is enforced by producing docile bodies within sites such as the prison, the school, the hospital, and the factory. The modern phase of political power arrives, for Foucault, when the production and regulation of the human as simple living being becomes what is at stake in a society's political arrangements. This symbiotic relationship between modern politics and biopower is announced in a famous passage in *The History of Sexuality*, volume 1:

> If one can apply the term *bio-history* to the pressures through which the movements of life and the procedures of history interfere with each other, one would have to speak of *bio-power* to designate what brought life and its mechanisms into the realm of explicit

calculation and made knowledge-power an agent of transformation of human life. It is not that life has been totally integrated into techniques that govern and administer it; it constantly escapes them. Outside the Western world famine exists, on a greater scale than ever; and the biological risks confronting the species are perhaps greater, and certainly more serious, than before the birth of microbiology. But what might be called a "threshold of modernity" has been reached when the life of the species is wagered on its own political strategies. For millennia, man remained what he was for Aristotle: an animal with the additional capacity for a political existence; modern man is an animal whose politics place his existence as a living being in question.[2]

Agamben accepts that a certain political threshold is reached when human life is wagered on the political strategies of a society, but his reading of what this threshold might be and what it means leads him to make two crucial modifications of Foucault's critique. He disputes, first of all, that the inscription of bare life (or *zoē*) within the *way of life* proper to an individual or group (*bios*) is a specifically modern phenomenon; such a maneuver, he argues, has *always* been at the basis of sovereign power and actually forms an unacknowledged point of connection between classical and modern thought. Agamben's identification of a type of continuum, rather than an abrupt biopolitical dislocation, between classical and modern politics leads him to a second modification of Foucault that forms the core of his argument. Instead of hypothesizing the relation between antiquity and modernity in terms of their maintenance or rejection of a biopolitical imperative, he suggests we need to recognize that the politics of both periods is determined by a "structure of the exception" (*HS*, 7). This structure of the exception is manifest in the Aristotelian definition of the *polis*, which is constituted around an opposition between life (*zen*) and good life (*eu zen*). However, this opposition is itself constituted by the simultaneous inclusion and exclusion of *bare life* within good life, since the former can be inscribed only within the latter at the price of its restriction within the domestic economy of the home and its segregation from politically qualified existence. The question that must be posed to classical and modern culture, Agamben maintains, is "why Western politics first constitutes itself through an exclusion (which is simultaneously an inclusion) of bare life" (*HS*, 7). From this perspective, the biopolitical imperative Foucault identifies within modernity is merely an extension of the task classical politics has always set itself—the politicization of bare life through a structure of exception that enables political power to distinguish between different forms of being:

> In assuming this task, modernity does nothing other than declare its own faithfulness to the essential structure of the metaphysical tradition. The fundamental categorial pair of Western politics is not that of friend/enemy but that of bare life/political existence, *zoē/bios*, exclusion/inclusion. There is politics because man is the living being who, in his own language, separates and opposes himself to his own bare life and, at the same time, maintains himself in relation to that bare life in an inclusive exclusion. (*HS*, 8)

The significance of the state of exception in Western culture, then, is that by simultaneously excluding bare life from, and capturing it within, the political order, it "actually constituted, in its very separateness, the hidden foundation on which the entire system rested" (*HS*, 9). The scope of this statement offers a productive general context in which to consider the function of life within the *polis*, but it becomes particularly compelling whenever we reflect on the relation between politics and law.

Carl Schmitt's definition of *sovereignty* ("Sovereign is he who decides on the state of exception") brings the relation between law, politics, and the state of exception forcefully into focus in its organization around two paradoxes.[3] Agamben's reading of Schmitt notes that the principal paradox of sovereignty "consists in the fact that the sovereign is, at the same time, outside and inside the juridical order" (*HS*, 15). The sovereign is always located *within* the law because the juridical principle is coextensive with his body and the field of his domination. Yet the sovereign can always also stand *outside* the law in order to suspend it and declare a state of exception or emergency. The classic example of this process, as cited by Agamben, is Auschwitz, where human bodies were once again declared merely to be biological material. Such a boundary could be redrawn only by a political power capable of shifting the very border between law and the human. The ambivalent position of sovereignty outside and inside the field of law is fundamental to the structure of exception insofar as it marks a threshold between interiority and exteriority that makes the concept of law *possible in the first place*. The internal consistency of the juridical principle is constituted by its relation to what it pushes outside itself in a form of exclusive inclusion glimpsed in the figure of the "ban" or the political exile who remains in legal subjection even as he or she is condemned to extrajuridical life. Indeed, the particular force of law, Agamben ruefully concedes, "consists in this capacity of law to maintain itself in relation to an exteriority" (*HS*, 18).

The paradox of the sovereign exception leads in turn to a related paradox, that of the nonjuridical origin of law. Schmitt elaborates on this paradox in *Political Theology*:

> The exception is that which cannot be subsumed; it defies general codification, but it simultaneously reveals a specifically juridical formal element: the decision in absolute purity. The exception appears in its absolute form when it is a question of creating a situation in which juridical rules can be valid. Every general rule demands a regular, everyday frame of life to which it can be factually applied and which is submitted to its regulations. . . . The decision reveals the essence of State authority most clearly. Here the decision must be distinguished from the juridical regulation, and (to formulate it paradoxically) authority proves itself not to need law to create law. . . . The exception is more interesting than the regular case. The latter proves nothing; the exception proves everything.[4]

As this passage makes clear, it is the structure of the exception rather than positive law or the example of the general case that creates the juridical principle. "For what is at issue in the sovereign exception," Agamben concludes, "is, according to Schmitt, the very condition of possibility of juridical rule and, along with it, the very meaning of State authority" (*HS*, 17).

Agamben provocatively suggests a disturbance has occurred within the juridico-political order in our own time that has profoundly altered the relation between politics and law. We can understand the nature of this disturbance by posing the following question: if, as Agamben claims, the modern politicization of bare life explored in Foucault reworks the "essential structure of the metaphysical tradition," how can we distinguish between the political ontology of classical and modern culture (*HS*, 18)? Agamben's response is to revise the Foucauldian thesis—what characterizes modern politics is the *inclusion* of bare life within the political realm—in order to demonstrate, instead, the "decisive fact" of political modernity: "together with the process by which the exception everywhere becomes the rule, the realm of bare life—which is originally situated at the margins of the political order—gradually begins to *coincide* with the political realm, and exclusion and inclusion, outside and inside, *bios* and *zoē*, right and fact, enter into a zone of irreducible indistinction" (*HS*, 9). Bare life is not merely included within the political order: it forms the *ontological ground* of modern politics. The reconstitution of bare life as the ontological ground of modern politics is evident in contemporary practices such as the industries of bioethics and population and health management, which enable the technocratic regulation of individual bodies and space fundamental to our governmental culture. This development, Agamben argues, produces in turn a shift in the function of the state of exception within political life that occurs when "the state of exception comes more and more to the fore as the fundamental political structure and begins to become the rule" (*HS*, 20). If juridico-political space is established by "a zone of irreducible indistinction" (the state of exception) between exclusion and inclusion, the transformation intrinsic to contemporary life is that the state of exception has become a topographical space that exists within, while also seeking to *replace*, the given political order. As noted earlier in *Homo Sacer*, Agamben identifies the archetypal example of this new exceptional principle as the concentration camp. What is distinctive about the camp in juridical terms is that it describes a space where the imbrication of the state of exception (based on the suspension of normative political logic) with law can be expressed. To be imprisoned in the camp is to be divested of any form of legal process or recourse, and this extrajuridical principle reciprocally constitutes the status of law in this new space. The distinction between exception and rule that underpins both classical and modern politics becomes inoperable because the camp introduces a "new juridico-political paradigm in which the norm becomes indistinguishable from the exception" (*HS*, 170). In the age of the camp, the meaning of both politics and law therefore undergoes a radical revision insofar as every action and statement now acquire significance only in their *abrogation* of normal legal process.

Writing Exception: The American Roman Noir

These reflections on Agamben, Schmitt, and the state of exception appear, at first glance, to be the proper object of cultural and political theory. In this essay, however, I want to suggest that they also offer the most productive theoretical context within which to understand the development of a particular *literary* genre: the American roman noir. What makes this genre of signal importance to a discussion of the structure of exception is its relentless interrogation of the relationship between politics and law in order to determine what might constitute coherent moral order in contemporary society. My analysis of this genre examines the ways in which it explores the relations among politics, exception, and law, but it does so to mark a fundamental shift in the function of the genre itself as it has developed over the last fifty years. It therefore begins with a reading of Raymond Chandler's noir classic *The Big Sleep*, arguing that Chandler's conservative political vision is sustained by its identification of a classical logic of exception within American modernity. The aesthetic politics of *The Big Sleep* is then juxtaposed with an analysis of James Ellroy's contemporary noir *L.A. Confidential*, demonstrating that the challenge posed by Ellroy's fiction consists in its transition from a classical to a modern structure of exception in which exception has now become the normative logic of juridical and political processes.

Perhaps the most powerful critique of the American roman noir as a form of aesthetic politics is found in the work of William Marling. Marling begins from the premise that American noir is a "dark style of narrative" that "emerged from the American imagination" as a reaction to the rapid economic and technological changes wrought by the 1920s.[5] This reaction, Marling insists, is a *moral* response: the narrative of roman noir expresses the judgment of social conscience on the excessive prodigality of the Jazz Age as it progressed inexorably toward the Wall Street crash and the Great Depression. Indeed, one of Marling's main claims is that roman noir is best understood as an obsessive reworking of the parable of the prodigal son insofar as its basic plot (or fabula) shifts among three points of view that highlight spending (the younger sibling), giving (the father), and saving (the older son). Above all, the parable of prodigality mapped out within roman noir is a narrative organized around the dichotomies of "inner/outer, near/far, soul/self, and desire/discipline" (*ARN*, x). Often these axes are misaligned, but noir narrative "works to restore harmony" in the terms prescribed by "familial discipline" and the demands of social order (*ARN*, xi).

The archetypal fabula of prodigality makes it, as Marling notes, a "wonderful economic as well as technological tool" (*ARN*, x). In order to establish the proper historical context for noir, we must see its appearance as contemporary with the economic shift from mass production (the example of Henry Ford) to mass marketing (the model of General Motors) and then to a culture of commodity consumerism (*ARN*, xi). What is often left unremarked, Marling reminds us, is that these economic and technological shifts were *also* shifts "from systems of desire to those of discipline" in which market forces and product design "evolved to promote and regularize consumption" (*ARN*, xii). The ambivalent position occupied by noir writers such as Chandler, Dashiell Hammett, and James M. Cain arises because they provide narrative forms by which the "perceptual revolution" produced by this economic revolution could be reexperienced in aesthetic terms. These narrative forms also cautioned about the consequences of the culture of commodity consumption from a moral and spiritual perspective. It is from this ambivalent perspective inside and outside the narrative of commodity consumption that Cain repeatedly highlights the conjoined perils of unregulated economic and sexual "spending" and Chandler advocates, in the figure of Philip Marlowe, the suave appeal of "the autoeroticism of self-discipline" (*ARN*, xii). Marling's reading of *The Big Sleep* as a "warning about the cost of financial and sexual prodigality" observes that while "the revealed story's outcome may be macabre or gloomy" the explicit function of noir narrative is "always reintegrative" (*ARN*, xiii). This reintegrative function can be formalized in the simple statement "desire must serve economic ends." It is in this sense that the narrative trajectory of the American roman noir is both conservative and corrective: "The American roman noir teaches us that things could be much darker. It prefigures our prodigality so we can imagine that we have avoided it" (*ARN*, xv).

Marling's analysis of roman noir works convincingly within the parameters it establishes, and it offers a useful socioeconomic context for the development of the genre. Its weakness, however, is that its location of economic critique within a discourse of moral improvement conceals the primary narrative business of roman noir, which is both a complex meditation on the way law is constituted and an organized politics that can articulate law and enforce divisions between different types of social being.

The concept of law in roman noir is determined by a structure of exception that creates an opposition between life and good life. This opposition turns on two axes: the figure of the detective outsider and the position of woman. The detective outsider is always excepted in noir fiction—his position as private detective simultaneously denotes an insufficiency within the established juridical order and his exclusion from the structure of institutional policing. At the same time, he paradoxically comes to guarantee the ethical potential of a society in which the relation between law and ethics has become obscure. Roman noir continually poses the question of the relation of society's outside to its inside and insists that what is excluded from the juridico-political order also represents the hidden foundation of that order. Meanwhile, the relation of exception to norm is also explored by means of a division within femininity or the principle of alterity with which the detective outsider must contend. Women in roman noir now inhabit a state of exception: they are both located within society—often high society— and routinely figured as existing outside its ethical and juridical protocols. However, the production of femininity as an exceptional principle is both endorsed and redefined by noir narrative, which reconfigures the exception of woman as a division *within and between* women that enables the genre to demarcate normative codes of behavior and social obligation.

It is a staple feature of detective fiction that the detective exists to bring order to a disordered environment. In contrast, however, *The Big Sleep* insists on the ambivalent position of the detective outside and inside the sphere of law. The novel wastes little time in informing us that Marlowe's talent for "insubordination" removed him from the police force and left him permanently at odds with policemen such as Captain Cronjager of the Los Angeles Police Department (LAPD).[6] Marlowe's refusal to compromise his professional obligation to his client, General Sternwood, despite his own involvement in a police inquiry, leads him into fierce dispute with Cronjager and the following exchange with District Attorney Taggart Wilde:

> Wilde tapped on his desk and stared at me with his clear blue eyes.
> "You ought to understand how any copper would feel about a cover-up like this," he said. "You'll have to make statements of all of it—at least for the files. I think it may be possible to keep the two killings separate and to keep General Sternwood's name out of both of them. Do you know why I'm not tearing your ear off?"
> "No. I expected to get both ears torn off."
> "What are you getting for it all?"
> "Twenty-five dollars a day plus expenses."
> "That would make fifty dollars and a little gasoline so far."
> "About that."
> He put his head on one side and rubbed the back of his left little finger along the lower edge of his chin.
> "And for that amount of money you're prepared to get yourself in Dutch with half the law enforcement of this county?"
> "I don't like it," I said. "But what the hell am I to do? I'm on a case. I'm selling what I have to sell to make a living. What little guts and intelligence the Lord gave me and a willingness to get pushed around in order to protect a client. It's against my principles to tell as much as

I've told tonight, without consulting the General. As for the cover-up, I've been in police business myself, as you know. They come a dime a dozen in any big city. Cops get very large and emphatic when an outsider tries to hide anything, but they do the same things themselves every other day, to oblige their friends or anyone with a little pull. And I'm not through. I'm still on the case. I'd do the same thing again, if I had to." (BS, 111–12)

Marlowe's principled iconoclasm is eventually formalized into a statement of ethical independence: "It's a question of professional pride. You know—professional pride" (BS, 152). Yet the remarkable feature of Chandler's fiction is its remorseless hostility to Marlowe's weary conclusion, "Once you're outside the law you're all the way outside" (BS, 187). In fact, Marlowe's ethical position outside the law is used to remind us of those values that once founded the law (and are still present in attenuated form in men such as Bernie Ohls and Taggart Wilde) but have now slipped into desuetude. This corrective function sometimes emerges as a defense of police procedure to those who, like Vivian Regan, simply conflate law with lawlessness. Thus, when Regan attempts to explain away Owen Taylor's police record by asserting that Taylor "didn't know the right people. That's all a police record means in this rotten crime-ridden country," Marlowe's tart rejoinder is "I wouldn't go that far" (BS, 59). The corrective function recurs more powerfully elsewhere, however, as implacable resistance to the identification of fact with law that underpins the blithe dogmatism of police business. Cronjager employs the term "police business" to describe an order of living and "thinking" rigorously distinct from the world of criminality and ordinary social transactions. Marlowe, by contrast, insists that the idea of law is preserved (and actually constituted) by calculated incursions into illegality:

Cronjager leaned slowly back in his chair and crossed one ankle over his knee and rubbed the ankle-bone with his thin nervous hand. His face wore a lean harsh frown. He said with deadly politeness:
"So all you did was not report a murder that happened last night and then spent today foxing around so that this kid of Geiger's could commit a second murder this evening."
"That's all," I said. "I was in a pretty tough spot. I guess I did wrong, but I wanted to protect my client and I hadn't any reason to think the boy would go gunning for Brody."

"That kind of thinking is police business, Marlowe. If Geiger's death had been reported last night, the books could never have been moved from the store to Brody's apartment. The kid would never have been led to Brody and wouldn't have killed him. Say Brody was living on borrowed time. His kind usually are. But a life is a life."

"Right," I said. "Tell that to your coppers next time they shoot down some scared petty larceny crook running away up an alley with a stolen spare." (*BS*, 107)

The theme of ethical independence is therefore crucial to Chandler's fiction because it is here that the contradiction in his view of law is expressed. As Marlowe's exchanges with Cronjager demonstrate, *The Big Sleep* continually highlights the exceptional structure of the juridical order while seeking at the same time to underline a distinction between the proper and improper uses of law. This distinction comes into force in the context of Marlowe's relation both to women and to economic class. Marlowe's dealings with the Sternwood family expose a crucial ambivalence in his attitude to social authority that resonates throughout the novel. This ambivalence becomes significant when we consider that his portrait of the Sternwoods provides a narrative of America's fall from the integrity of its origins into a culture of commodification and spectacle. The belief in a fall from an ideal social order is suggested on the opening page of the novel in which Marlowe arrives at the Sternwood home to be confronted with a stained-glass panel depicting "a Knight in dark armor rescuing a lady who was tied to a tree and didn't have any clothes on but some very long and convenient hair." The knight, however, had "pushed the visor of his helmet back to be sociable," was "not getting anywhere" in his attempted rescue, and, in fact, "didn't seem to be really trying." Alone outside the house, Marlowe is struck by the melancholy thought, "If I lived in the house, I would sooner or later have to climb up there and help him" (*BS*, 9).

This scene is significant for two reasons. It offers the first indication of a dislocation between a mythic past and a decadent present that will compel Marlowe to articulate his own moral code outside the values represented by the dominant social class. Much later in the novel, following a confrontation with Sternwood's daughter Carmen, he is forced to admit that "Knights had no meaning in this game. It wasn't a game for Knights" (*BS*, 153). The opening scene is also notable for identifying a schism between the origin and function of wealth that helps to structure the novel's moral economy. This schism is expressed in terms of the difference in social attitude between dying General Sternwood and his daughters, Vivian and Carmen. The novel's extraordinary opening scenes depict the general as a creature artificially kept alive in a society that he can no longer recognize. Wheeled by his butler into "a sort of vestibule that was about as warm as a slow oven," his crippled body languishes while a "few locks of white hair clung to his scalp, like wild flowers fighting for life on a bare rock" (*BS*, 13–14). These images of declension highlight a contradiction between Sternwood's rhetoric and the material conditions that underpin his moral code and social position. Set apart from 1930s society by his contempt for its mercantile ethic, he is shown drinking "champagne as cold as Valley Forge" (*BS*, 14) (a reference to the scene of George Washington's retreat during the American Revolution). At the end of the novel, he will question Marlowe's "ethical" professional conduct because the structure of ethical obligation, in his view, forms the ground of social relations (*BS*, 203). However, Sternwood's social position is actually maintained by his role as an oil baron, and oil becomes a metaphor in the novel for the transformation of material forces into forms of social division:

> I stood on the steps breathing my cigarette smoke and looking down a succession of terraces with flowerbeds and trimmed trees to the high iron fence with gilt spears that hemmed in the estate. A winding driveway dropped down between retaining walls to the open iron gates. Beyond the fence the hill sloped for several miles. On this lower level faint and far off I could just barely see some of the old wooden derricks of the oilfield from which the Sternwoods had made their money. Most of the field was public park now, cleaned up and donated to the city by General Sternwood. But a little of it was still producing in groups of wells pumping five or six barrels a day. The Sternwoods, having moved up the hill, could no longer smell the stale water sump or the oil, but they could still look out of their front windows and see what had made them rich. If they wanted to. I didn't think they would want to. (*BS*, 25–26)

"To hell with the rich," Marlowe remarks at one point, "they make me sick" (*BS*, 66). This statement is often taken as a definitive judgment on the Sternwood family. Paul Skenazy makes this point when reading the novel as a critique of the "insularity of money."[7] The difficulty in reading the novel in these terms, however, is that Marlowe repeatedly distinguishes *between* General Sternwood and his daughters in his attitude toward socioeconomic class. The wasted body of General Sternwood does not represent to Marlowe a lifetime of hedonistic self-gratification; instead, he looks on it as a symbol for the disappearance of a vital connection between economic and moral autonomy. It is Sternwood, after all, who cleans up and donates land to the city, who spends his last days attempting to preserve the reputation of his daughters, and with whom Marlowe enters into his most extended discussion of professional ethics at the end of the novel. His daughters, in contrast, have absolutely no conception of the social contexts in which capital operates, and this is demonstrated by their enthrallment to the spectacle of images and commodities. Vivian spends her evenings running up gambling debts she has no intention of paying (a transaction in which money is twice devalued, as roulette chip and as canceled promissory note), while Marlowe's entry into the Sternwood sphere is precipitated by Carmen's involvement in a pornography racket (the body transformed into a type of capitalized image), which leads to the death of Arthur Geiger. It is precisely this transformation of the social context of wealth into an empty self-reflexive space in which the effects of capital dominate individual consciousness that is captured in the description of Vivian's bedroom:

> This room was too big, the ceiling was too high, the doors were too tall, and the white carpet that went from wall to wall looked like a fresh fall of snow at Lake Arrowhead. There were full-length mirrors and crystal doodads all over the place. The ivory furniture had chromium on it, and the enormous ivory drapes lay tumbled on the white carpet a yard from the windows. The white made the ivory look dirty and the ivory made the white look bled out. The windows stared towards the darkening foothills. It was going to rain soon. There was pressure in the air already. (*BS*, 22)

The moral distinction Marlowe's narrative asserts between wealth creation and commodity consumption is reinforced by his prejudice against improper sexual "spending." To underline this point, *The Big Sleep* enforces a division *within* the feminine between two types of women. Vivian, it is true, is hardly a model of sexual propriety (and her casual extravagance is exemplified by her three marriages), but she is also compelled to adopt a maternal role toward her erratic younger sister. Marlowe's portrait of Carmen produces an image of aberrant femininity that exists almost wholly outside social roles or any recognizable set of social relations. In his interview with Marlowe at the beginning of the novel, General Sternwood remarks that the perfume of orchids has "the rotten sweetness of a prostitute" (*BS*, 14), and this association of female sexuality, corruption, and illegality is displaced onto Carmen as the narrative proceeds. But Carmen exists not just beyond the boundaries of sexual propriety; her sexual "excess," which is constantly represented in her discourse, her dress, and her propensity to appear naked in strange men's apartments, takes her in turn beyond the boundaries of the *human*. She is described from the earliest moments as the hybrid of an animal and an erotic automaton: "She came over near me," Marlowe recounts, "and smiled with her mouth and she had little sharp predatory teeth, as white as fresh orange pith and as shiny as porcelain" (*BS*, 10).

In her discussion of the mytheme of the "dangerous" woman in (film) noir narrative, Janey Place suggests that the "ideological operation of the myth (the absolute necessity of controlling the strong, sexual woman) is thus achieved by first demonstrating her dangerous power and its frightening power, then destroying it."[8] Marlowe's description of Carmen conforms rigorously to this paradigm: she represents, in fact, the danger of unregulated desire or the uncoupling of libidinal investment from a proper social object. The price Carmen pays for this transgression is for her to be judged insane, and the novel concludes with Marlowe hoping that a "cure" might be found for the erotomania that led her to murder Rusty Regan for refusing her advances. This judgment is not passed, however, before Chandler underlines once more the connection between the decadence of modern Los Angeles and the "problem" of unregulated female desire. In the final, set piece scene of the novel, Carmen lures Marlowe to the abandoned oil field where she shot Regan. Apart from Carmen's unsuccessful murder attempt—Marlowe gives her a gun containing blanks—this scene records the complete destruction of her as a human being. This destruction is described in orgiastic terms: Carmen bares her "sharp little teeth" and begins to "hiss"; her face resembles an "animal, and not a nice animal"; eventually "her whole face [goes] to pieces" as she slips beyond consciousness, only to wake and giggle, "I wet myself" (*BS*, 210–11). Her orgiastic abandonment is also an act of historical desecration because it takes place where "the oil-stained, motionless walking beam of a squat wooden derrick" signifies the place where the Sternwood fortune was made. Libidinal flow is now continuous with the wasting of an inheritance as Marlowe recoils from the "stagnant, oil-scummed water of an old sump iridescent in the sunlight" (*BS*, 209).

Marlowe's curiously doubled position inside and outside the juridical sphere is constituted through this confrontation with unregulated economic-libidinal flow. His role in Chandler's fiction is ceaselessly to mark a threshold between the outside and inside of law that enables juridical and social order to be articulated. It is no surprise that it is when he is in the company of Carmen that Marlowe realizes the inherited codes of masculine "chivalry" are outmoded in the modern world. His physical and verbal control of Carmen permits him to project an image of secure and self-aware masculinity to counter those moments that suggest the origins of his rigid moral code lie in his repression of a homoerotic investment in other socially marginalized figures. (One thinks of his description of the "very handsome" homosexual Carol Lundgren's "dark eyes shaped like almonds, and a pallid handsome face with wavy black hair growing low on the forehead in two points," and his lyrical lament for Harry Jones, which is spoken out loud "in a voice that sounded queer to me" [BS, 97, 173].) Elsewhere Marlowe's relationship with General Sternwood underlines his identification with the pioneer spirit of self-determination that has become submerged within the culture of commodity consumption but that sustains the ethical rigor of his private code. Just as the narrative trajectory of classical detective fiction works backward from the scene of a crime to the circumstances that explain it, so Chandler's roman noir expresses American modernity by its exception of the detective outsider reconfigured as a type of Jeffersonian husbandman. This structure of exception forms the ground for every conflict within noir narrative, and it is these divisions that film noir represents so forcefully to the popular imagination.

The Law as Exception: *L.A. Confidential*

At first glance, the fiction of contemporary American crime writer James Ellroy appears to be situated securely within the traditions of roman noir popularized by Chandler. For if Chandler's novels describe a structure of exception in which what is *marginalized*—the values embodied by the detective outsider—is precisely what should be *normative* within its juridical and political institutions while what is normalized is actually fatal to any sense of legal and ethical propriety, we can also observe this conflict between normativity and aberration at the core of *L.A. Confidential*, the third installment of Ellroy's celebrated *L.A. Quartet*. If a novel so various in its stylistic effects and narrative trajectories may be said to have a central theme, it is surely what constitutes juridical order in a society where legality and criminality have become indistinguishable. This theme is explored at every level of the novel. It is dramatized in the struggle of Detective Edmund Exley to resolve the mystery of the "Nite Owl" murders, which leads him to uncover, among other things, both a conspiracy within the LAPD to seize control of the Los Angeles drug and pornography market and the murderous past of his own father, Preston Exley, a former LAPD detective. It recurs in the tension the novel identifies between Exley's rhetoric of "absolute justice" and his fraudulent and politically motivated reconstruction of his own biography. It also finds an enduring symbol in the figure of Lynn Bracken, the call girl made to look like Veronica Lake, who exists simultaneously within and without high society—and, crucially, within and without any sense of what it means to have a continuous relationship to one's own identity or values—and whose function it is partly to force both male protagonists, Exley and Bud White, to come to terms with the "dark places" within them, which they use the rhetoric of justice to obscure.

Ellroy's *L.A. Quartet* offers, however, a radically different vision of law from Chandler's roman noir. What is remarkable about Ellroy's fiction is that it combines an almost insatiable appetite for encoding and ironizing noir motifs and conventions while refusing to mark or respect the moral division between normativity and aberration (or the outside and inside of the juridical sphere) that noir repeatedly invokes. Ellroy's refusal to provide an appropriate moral context within which we can read signs as deviant or normative is in part what makes his work difficult and challenging. Another reason for the distinctiveness of Ellroy's fiction is its identification of the modern structure of exception at the heart of American law. In contrast to *The Big Sleep*, which examines the ambivalent position of the detective within and without juridical structures, *L.A. Confidential* describes a world in which the exception now coincides absolutely with the rule. Marlowe's unrequited desire for a separation between juridical and extrajuridical order is unsustainable in Ellroy's work, which insists that the idea of law is produced by the *suspension* of the juridical principle. The rule of law in *L.A. Confidential* is, in fact, the effect of a generalized state of emergency in which the distinction between norm and aberration is no longer operative.

The impossibility of distinguishing between norm and aberration in *L.A. Confidential* can be established insofar as the characters that represent or enforce "normative" values either are always already exposed as compromised in their relation to the values they espouse or are compelled to betray those values in the act of asserting them. White's entire police career is motivated by a desire to avenge his mother's violent death at the hands of his father and to punish men who beat women, yet he beats Bracken to punish her for her affair with Ed Exley. Preston Exley and Raymond Dieterling construct the Disney-style space of Dream-a-Dreamland as a monument to the potential for creative renewal inherent in the American way of life, and yet this dream of self-transformation is indissociable in the novel from the "dreaming" of the killer David Mertens, Dieterling's illegitimate son, who murders and dismembers children with his accomplice Loren Atherton in the hope of "creating children to their own specification" and re-creating the real as a pornographic dystopia.[9] Meanwhile, Ed Exley's "outsider" status in the LAPD is defined by his principled dependence on fact and evidence; but his reputation and extraordinary rise through the police hierarchy are underpinned by his fraudulent claim to have heroically wiped out a group of Japanese soldiers in a World War II gunfight.

The paradoxical status of law in *L.A. Confidential* as a permanently enforced suspension of juridical rule is confirmed in an early scene in which Preston Exley, chief of Exley Construction, compares his son Ed to his dead son Thomas and questions Ed's suitability to be an LAPD detective. It becomes clear, as the exchange progresses, that any legal value that might exist outside criminality against which the category of the criminal might be established is always already present *within* the criminal as a form of aberrant practice:

> "Father, Thomas was going to be your chief of detectives, but he's dead. Don't deny me my opportunity. Don't make me live an old dream of yours."
>
> Preston stared at his son. "Point taken, and I commend you for speaking up. And granted, that was my original dream. But the truth is that I don't think you have the eye for human weakness that makes a good detective."
>
> His brother: a math brain crazed for pretty girls. "And Thomas did?"
>
> "Yes."
>
> "Father, I would have shot that purse snatcher the second he went for his pocket."
>
> De Spain said, "Goddammit"; Preston shushed him. "That's all right. Edmund, a few questions before I return to my guests. One,

would you be willing to plant corroborative evidence on a suspect you knew was guilty in order to ensure an indictment?"

"I'd have to—"

"Answer yes or no."

"I . . . no."

"Would you be willing to shoot hardened armed robbers in the back to offset the chance that they might utilize flaws in the legal system and go free?"

"I . . ."

"Yes or no, Edmund."

"No."

"And would you be willing to beat confessions out of suspects you knew to be guilty?"

"No."

"Would you be willing to rig crime scene evidence to support a prosecuting attorney's working hypothesis?"

"No."

Preston sighed. "Then for God's sake, stick to assignments where you won't have to make those choices. Use the superior intelligence the good Lord gave you." (*LAC*, 19–20)

From this point illegality is established as the condition of juridical order. The novel quashes any appeal to an idealized form of law in the prologue, in which LAPD detective Dudley Smith murders former policeman Buzz Meeks for a cache of heroin stolen from incarcerated mobster Mickey Cohen. The traditional paradigm of roman noir, which effects a moral separation between the detective outsider, on one hand, and criminality and corrupt policemen, on the other, is replaced by a structure in which legality and illegality are retrospective fictions appended to a common set of motivations. The irony of *L.A. Confidential* is that the 1940s, the period in which Cohen and his associates controlled the Los Angeles drug market, appears from the perspective of the early 1950s as a tranquil golden age when the distinction between crime and law retained some meaning. To say this is not simply to assert that law, in the hands of Dudley Smith or Preston Exley, has become a form of legalized crime, but that there now appears to be no position within the law from which disorder can be identified. The seamless continuity between law and crime is perhaps represented most dramatically in the figure of Los Angeles District Attorney Ellis Loew, the senior legal figure in the city, whose career is sustained by Smith's drug money and the sex killer Spade Cooley's "slush fund," and who is personally implicated in the pornography racket organized by Smith and Pierce Patchett (*LAC*, 464).

Nowhere is the disappearance of a threshold between law and crime more vividly illustrated, though, than in the career of Ed Exley, who begins the novel with a naive faith in his father's commitment to "the solving of crimes that require absolute justice" and concludes it by securing White's support against Smith and Cooley with the promise, "I'll let you kill them" (*LAC*, 19, 429). Exley is crucial to *L.A. Confidential* in symbolic, as well as narrative, terms because he demonstrates that the rhetoric of justice does not betoken fidelity to a set of abstract principles; instead, it enables characters to invest in a *style of morality* in order that others will cede them authority within the environment they inhabit. Exley's rhetorical commitment to law as a mode of self-formation is exposed by Inez de Soto, the rape victim with whom he begins a sexual relationship, who tells him that "the *big* lie" is "you and your precious absolute justice" (*LAC*, 275). The disintegration of the juridical as a determining principle is underlined at the end of the novel in an exchange between Exley and Gallaudet:

> Downtown to the Dining Car: a bright place full of nice people. Gallaudet at the bar, sipping a martini. "Bad news on Dudley. You don't want to hear this."
>
> "It can't be any worse than some other things I've heard today."
>
> "Yeah? Well, Dudley's scot-free. Lana Turner's daughter just knifed Johnny Stompanato. D. O. fucking A. Fisk was staked out across the street and saw the meat wagon and the Beverly Hills P. D. take Johnny away. No Dudley witness, no Dudley evidence. Grand, lad."
>
> Ed grabbed the martini, killed it. "Fuck Dudley sideways. I've got a shitload of Patchett's money for a bankroll, and I'll burn down that Irish cocksucker if it's the last thing I ever do. *Lad.*"
>
> Gallaudet laughed. "May I make an observation, Inspector?"
>
> "Sure."
>
> "You sound more like Bud White every day." (*LAC*, 471–72)

The loss of the juridical as an ethical ground on which different categories of behavior and motivation could be established is exacerbated by Ellroy's strategy of drawing together within a relation of contamination ways of structuring the "real" that we feel should be rigorously separated. This strategy is evident in the comparison Ellroy draws between police business and the hyperreal conventions of Hollywood cinema, television programming, and journalism. The reciprocal relationship between the worlds of law and entertainment is underlined by the activities of "Hollywood" Jack Vincennes, a detective whose career has stalled and who now makes his living (and garners his self-esteem) from his role as "police advisor" to the television show *Badge of Honor* and the information he supplies to Sid Hudgens, proprietor of Hollywood scandal sheet *Hush-Hush*. Vincennes is pivotal to *L.A. Confidential* because through his character Ellroy makes the connection between the fabrication of television plots, scenarios, and "scenes" and the hyperreal construction of legality—through "plot devices" such as the planting of evidence, the preparation of false statements, and the use of newspaper reports to undermine defense attorneys—within police practice.[10] "I'm a cop and I'm Hollywood" (*LAC*, 226), Vincennes retorts at one point to an inquiry about his professional "duty," and his position in the novel, at least in its early stages, is the ironic complement to Sid Hudgens, who invents every detail he cannot establish while cleaving to the rhetoric of professional detective work ("But let's talk facts" [*LAC*, 88]).

Conclusion: Ellroy and the Politics of Style

If we may define *style* as the mode of relation between thought and its representation, Ellroy's continual subversion of the "proper" relation between event and context foregrounds the problem of how order is constituted in modern American society. The question of Ellroy's literary style is something each of his readers must confront, but it is an issue that has received little discussion beyond journalistic references to its "energy" or "staccato" quality.[11] A more profitable approach to Ellroy's style, one might argue, would be to displace it from the noir tradition in which it is usually considered and read it instead in the context of the so-called paranoid school of American writing exemplified by figures such as Thomas Pynchon, Ishmael Reed, and Don DeLillo. The following paragraph is typical of the style of *L.A. Confidential*:

> Heroin and pornography lines. "The Guy" who made the smut books as Sid Hudgens's killer, his front man Duke Cathcart—killed by Dean Van Gelder, ordered killed or merely approached by Davey Goldman—who learned of the smut proposal via the bug in Mickey Cohen's cell. Cohen omnipresent—his stolen heroin ended up with both the Engleklings and "The Man" who brought Patchett the eighteen pounds of

"H" for development, "The Man" who also loved pornography and convinced Patchett to manufacture new books from the 1953 prototypes. An instinct: Cohen was Mr. Patsy going back eight years, in and out of jail, a focal point who never dealt his own hand into the welter of the cases. A line to a conclusion: the Nite Owl killings were semiprofessional at least, an attempt to take over the heroin and pornography rackets of Pierce Patchett. Cathcart, attempting to push the smut on his own, was the focus of the killings. Did he misrepresent his importance to the wrong people, or did the shooters deliberately take out Van Gelder, knowing or not knowing he was a Cathcart impersonator? (*LAC*, 396)

It is clear from an extract like this that, unlike roman and film noir, Ellroy's style is not principally distinguished by its formal control of indexical narrative functions like tone and atmosphere. The genre of noir is compelled to privilege indexical functions because it employs the vocabulary of aesthetic "darkness" to highlight problems of moral order; it is this relation between aesthetic and moral darkness that the chiaroscuro lighting effects of film noir seek to establish.[12] Ellroy's style, on the other hand, produces problems at the level of *structure* rather than tone because it proliferates plotlines, narratives, and types of character beyond any principle of comprehension or readability that we can master. Another way of putting this is that his fiction inculcates a principle of calculated excess: it always provides the reader with far too much information with which to order the narrative world(s) into coherence, and it does so in order to dramatize Ellroy's conviction that the privatization of power in modern America is predicated on a ruthless division between private and public knowledge. The law in 1950s Los Angeles is in the hands of men like Dudley Smith, and we are powerless before them precisely because the multiplicity of narrative contexts they bring to bear on discrete forms of knowledge exceeds the capacity (or the point of access to information) of the reader as private citizen.

Ellroy's fiction undoubtedly shares many of the thematic and formal preoccupations of the paranoid style, and to place it within this tradition opens it to a political reading beyond the confines of crime writing. What must also be acknowledged, however, is that his radicalization of narrative structure deliberately undermines the possibility of a metanarrative context outside the various textual discourses that might offer a normative standard against which types of rhetoric or forms of behavior could be understood as deviant or transgressive. Much of the difficulty of reading *L.A. Confidential* arises from Ellroy's concerted use of free indirect style to create a kind of rhetoric indifferently positioned between first- and third-person narrative. The effect of this technique is to blur any distinction between authorial statement and character motivation, so that it becomes impossible to determine what constitutes normative or exceptional behavior in any particular context. This difficulty in distinguishing between norm and exception is vividly illustrated in the novel's treatment of race. *L.A. Confidential* is saturated in the rhetoric of racism: the distinction between who is included or excluded from ethical consideration within the *polis* is underlined by Hudgens's free-form riffs on the "dusky deelites" who provoke "dark desires" down in "darktown," where Dudley Smith organizes a militia to repress the social aspirations of blacks (*LAC*, 89). Ellroy's novel, however, provides us with no position from which to read this racist rhetoric *as* exceptional because its omniscient third-person narrative *reinscribes* it in a series of narrative judgments. In one of the key scenes of the novel, Bud White storms an apartment in "darktown" and shoots the man who has imprisoned and raped Inez de Soto. "Bud shot him in the face, pulled a spare piece," the narrator informs us, "bang bang from the coon's line of fire" (*LAC*, 133). It is not simply the casual and extrajuridical execution of a criminal suspect that is shocking here but the recognition that the term *extrajuridical* retains no legitimate meaning. What lies outside the law now constitutes the law, and Ellroy's style registers the effect of the modern state of exception on our legal and ethical protocols with exemplary power.

    The originality of Ellroy's fiction lies in the challenge it presents to the way we conceive of the relation between politics and law. No doubt the location of his work in 1940s and 1950s Los Angeles conditions the popular perception of him as a writer faithful to the tradition of American roman noir, but his writing continually reworks the structure of the juridico-political sphere that noir sought to interrogate. Where Chandler's novels attempt to realign politics and law with a moral code now in abeyance, *L.A. Confidential* explores the way politics *comes into being* as a relation of power by establishing an outside and a state of exception that reciprocally constitutes the interior epistemological space of law. In Ellroy's other work, the outside of law that establishes the space of law and the form of political rhetoric in American culture is frequently figured as the murdered and degraded body of a woman—one thinks of Elizabeth Short in *The Black Dahlia* and Ellroy's mother Geneva "Jean" Hilliker Ellroy in *My Dark Places*—which extends the possibility of a gendered critique of power relations in his reading of law. These readings remain to come, but they will do so in the knowledge that it is his vision of law and the state of exception—of law *as* the state of exception—that is the risk and the reward of Ellroy's writing.

Notes

1. Giorgio Agamben, *Homo Sacer: Sovereign Power and Bare Life*, trans. Daniel Heller-Roazen (Stanford, CA: Stanford University Press, 1998), 2. Hereafter cited parenthetically by page number as *HS*.
2. Michel Foucault, *The History of Sexuality*, vol. 1, *An Introduction*, trans. Robert Hurley (London: Penguin, 1990), 143.
3. Carl Schmitt, *Political Theology: Four Chapters on the Concept of Sovereignty*, trans. George Schwab (Cambridge, MA: MIT Press, 1985), 5.
4. Schmitt, *Political Theology*, 13.
5. William Marling, *The American Roman Noir: Hammett, Cain, and Chandler* (Athens: University of Georgia Press, 1995), ix. Hereafter cited parenthetically by page number as *ARN*.
6. Raymond Chandler, *The Big Sleep* (1939; London: Penguin Books, 1948), 15. Hereafter cited parenthetically as *BS*.
7. Paul Skenazy, *The New Wild West: The Urban Mysteries of Dashiell Hammett and Raymond Chandler* (Boise, ID: Boise State University Press, 1982), 40.
8. Janey Place, "Women in Film Noir," in *Women in Film Noir*, ed. E. Ann Kaplan (London: British Film Institute, 1988), 56.
9. James Ellroy, *L.A. Confidential* (London: Arrow Books, 1994), 466.
10. Ellroy's emphasis on the hyperreal construction of legality within police practice is taken one step further in Curtis Hanson's 1998 film version of *L.A. Confidential*. Hanson's commitment to the reconstruction of the real as artifice is evidenced by his decision to open the movie with the voice of Sid Hudgens, who represents inauthenticity itself in the novel. This theme is particularly hypertrophied in Hanson's treatment of the character of Lynn Bracken, where the visual quality of cinema allows Hanson to take the problem of the hyperreal further than Ellroy. The criminal rationale for the Fleur-de-Lis prostitution and extortion racket was that of prostitutes surgically altered to look like movie stars: fantasy and reality became indistinguishable as one had sex with the movie star of one's choice. In this world, Bracken was made to look exactly like Veronica Lake. What is interesting in the film, though, is that Hanson casts as Bracken an actress (Kim Basinger) who herself has an iconic sex symbol status within a different Hollywood economy. In this sense the viewer has to follow the movement of Bracken-Lake-Basinger through a potentially endless series of metonymies in which what is real and what is simulated is impossible to determine and the question, "So does Basinger actually *look* like Lake?" is wholly beside the point. This vertiginous movement between simulation and the "real" is reproduced, of course, in one of the film's finest comic moments when Ed Exley sneers at Johnny Stompanato's female companion for being made to look like Lana Turner, only to be informed that the subject of his derision is Lana Turner herself.

11 A typical example of this tendency is on the dust jacket of the Arrow edition of *L.A. Confidential*: "Ellroy writes as if driven by demons. His brutal, staccato graffiti tips over into art."
12 The clearest explanation of the term *indices* is given by Roland Barthes in "Structural Analysis of Narratives." For Barthes, indices are "integrational narrative units" that do not, unlike distributional units, denote narrative functions on the same level (picking up a telephone having as its correlate the moment when it will be put down) but that rather refer "to a more or less diffuse concept which is nevertheless necessary to the meaning of the story: psychological indices concerning the characters, data regarding their identity, notions of 'atmosphere,' and so on." Barthes, "Structural Analysis of Narratives," *Image-Music-Text*, trans. Stephen Heath (London: Fontana, 1977), 92.

# Adrian Mackenzie

Suspended Animation: Thinking and Animality in Neurocultural Selfhood

Everyday events sometimes call our sense of selfhood into question. They trigger reactions and behaviors that have more to do with territory, survival, and animal communication than with meaning, intention, identity, or thought. Reactions of fear, flight, or attachment can occur anytime: at a political event, in the theater or art gallery, at work, or on the street. This essay analyzes a strange form of self-improvement literature animated by such events. The literature assembles knowledges and techniques of selfhood based on scientific models of animal behavior, cognition, and physiology. Many forms, practices, and ideas of communication, sociality, values, law, ethics, and technology blur the lines between humanity and animality. Many examples in different domains could be cited here, but all of them transform and represent the everyday experience of living as a set of responses to be analyzed, monitored, and regulated through cognitive, behavioral, and neurophysiological models of animality. Giorgio Agamben's work allows one thread of this development to be unraveled and evaluated. Agamben analyzes how concepts of life underpin political formations and forms of power. Crucially, he frames *thinking* as a form-of-life. This framing provides ways of situating animalization in relation to thinking (as well as in relation to responsibility, ethics, politics, and futurity). Since thinking or thought retains a special privilege in animalized accounts of personhood, self, and relation to others, the connections that Agamben makes between thinking and life have special importance. Any shift in thinking about thought or in practices of thinking deeply affects experiences of self, body, others, and collective life. Agamben's work demonstrates why affirming animality remains difficult for those who like to think of themselves as human. It articulates an important question: in what way can we become animals?

Animalization as Everyday Experimental Sensibility

The literature analyzed in this essay forms a loose corpus situated at the boundary between books on popular science, personal development, and cultural theory. In his best-selling book *Blink: The Power of Thinking without Thinking*, Malcolm Gladwell asks, "What would happen if we if took our instincts seriously?"[1] Taking instincts seriously in itself is not new. Since the nineteenth century, large parts of the psy-disciplines (psychology, psychoanalysis, cognitive science) and the life sciences have been devoted to that. Gladwell answers that taking "our instincts seriously" means attending to "the very smallest components of our everyday lives—the content and origin of those instantaneous impressions and conclusions that spontaneously arise whenever we meet a new person, confront a complex situation or have to make a decision under conditions of stress" (*B*, 16). We can take this call to heed our instincts in different ways. For instance, we might regard it as reaffirming biological determinism. However, Gladwell and others situate biology and animality in everyday senses of self. They highlight everyday scenarios (in the classroom, on the street, in a gallery, at home) and suggest that understanding biological explanations of them changes selfhood. In *Mind Wide Open: Why You Are What You Think*, another recent best-selling nonfiction book on thinking, Steven Johnson suggests, "Knowing something about the brain's mechanics—and particularly *your* brain's mechanics—widens your self-awareness as powerfully as any therapy or meditation or drug. Brain science has become an avenue for introspection, a way of bridging the physiological reality of your brain with the mental life you already inhabit."[2] There is much to analyze here. Almost every word in this brief passage—*knowing, brain mechanics, self-awareness,* and so on—opens onto vistas of debate, contention, politicized struggle, and power. The political theorist William Connolly goes even farther by saying, "Thinking is neurocultural."[3]

Rather than being concerned with neurophysiology or cognitive psychology as scientific enterprises, Gladwell, Johnson, and Connolly draw from those fields practically oriented ideas about selfhood in everyday life. They look for elements to weave together in "technologies of self"[4] that can rapidly react, feel, invent, intuit, and organize themselves in networked capitalism. Johnson, like many others, sees brain science as a way of delineating how thinking overflows the discursive, logical, rational, or representational. The connection to brain science affirmed here is not to a classic scientific model or to a determinist or reductionist account of self or culture, as might be found in contemporary evolutionary psychology.[5] These accounts inculcate a quasi-experimental relation to self by borrowing from specific scientific disciplines and knowledges, combined with television nature documentary–style narrative, brain-imaging techniques, therapeutic interventions, and dietary and pharmaceutical regimens. Media, technology, everyday life, and science intermingle here. Connolly connects the different components in this experimental sensibility when he writes, "Today a dense series of loops and counterloops among cinema, TV, philosophy, neurophysiology, and everyday life enable people to explore the relation between thinking and affect more readily" (N, 67). While the precise character of the "loops and counterloops" among media, science, and everyday life merit more description, here the key problem is to explain how animalization makes these loops feasible.

Animality in Everyday Action

At one level, Agamben's *The Open* parallels the literature discussed above.[6] Johnson, Gladwell, and Connolly (although Connolly's account diverges radically in theory) connect thinking and animality at a practical, quasi-popular level. They combine science and quotidian anecdotes to prove that responses, feelings, intuitions, or various kinds of memory are animal reactions developed to solve problems of survival. Agamben's work also connects thinking to animality. However, it scales up thinking to a historico-philosophical and political level. The practical, quasi-popular level understands thinking as something that can be modified, enhanced, and rendered more flexible or adaptable by reference to scientific knowledges of animal behavior, cognition, and neurophysiology. As Nigel Thrift argues, these modifications serve very precise economic functions.[7] Similarly, at the historico-philosophical level of Agamben's account, the machinery that produces the very possibility of any experience of boredom, pleasure, everyday life, and thinking hinges on animality, as we will see. However, Agamben's work differs decisively. It allows us to ask whether the loops between neurophysiology and everyday life, between animality and thinking more broadly, can be smooth or uneventful.

In *The Open*, Agamben argues that any concept of humanity must both exclude *and* include animal nature. A logic of exclusion-inclusion drives the history of philosophical, religious, legal, political, scientific, and artistic concepts of the human in the West. Departing from an analysis of political sovereignty developed in *Homo Sacer*,[8] Agamben formulates this logic as the "anthropological machine" (O, 80). This mechanism generates figures of the human by dividing or cutting between human and animal. Yet, at each moment, the anthropological machine teeters on the verge of breakdown. "The motor of the historical becoming of the human" has repeatedly produced humanity in tension with animality (O, 80).

How does the anthropological machine produce any historical becoming of the human? Like the machinery of Kafka's penal colony, the machine that Agamben describes performs a series of topological operations on the living. It cuts living elements apart (for instance, body and soul) *and* folds this cut back into the living in order to produce the human. It divides animality from ideas, practices, and discourses of being human, and yet, it also includes animality within them. Because the anthropological machine constantly reinscribes animal-human difference within the human, figures of humanity cannot stabilize. The machine intermittently injects new breaks and separations. In the history of Western ideas of the living, Aristotle, Thomas Aquinas, Carolus Linnaeus, and Ernst Haeckel represent significant philosophical and scientific articulations of the same discursive machinery. The different versions of the exclusion-inclusion share the folded topology:

> Inasmuch as the production of the human by the opposition man/animal, human/inhuman is in play in it, the machine functions necessarily by an exclusion (which is also and always already a capture) and an inclusion (which is always and already an exclusion). It is precisely because the human is, in effect, each time already presupposed that the machine produces in reality a kind of state of exception, a zone of

indetermination where the outside is only the exclusion of the inside and the inside, in its turn, only an exclusion of the outside. (*O*, 37; translation modified)

As in *Homo Sacer*, Agamben calls this space of included-exclusion a "state of exception." In both older and newer versions, the anthropological machine generates a "zone of indetermination," or state of exception, in which human and animal remain or become indistinguishable. The zone takes different forms, but it always posits the existence of something not yet human that is already human. For instance, late-nineteenth-century accounts of human evolution solved the problem of the "missing link" by envisaging the virtual existence of prelinguistic humans.

No one lives only in the zone of indetermination. The state of exception precipitates from incessant cleaving, rearticulating, dislocating, and displacing of human and animal lives. Given the topological kinks of the anthropological construct, there is no prospect of simply putting humanity and animality back together again. Rather than trying to reunite these elements or substances (as have many philosophical projects during the last centuries and as does the literature under analysis here), Agamben seeks to extract the dynamics of their constant division: "We must, on the contrary, learn to think man [*sic*] as what results from the disconnection of these two elements [body and soul] and examine not the metaphysical mystery of their conjunction, but the practical and political mystery of the separation" (*O*, 16; translation modified). Practical and political decouplings or disconnections of animal bodies and human souls produce the human. In whatever sense, human life reiterates that dividing and breaking apart, more or less incessantly, in many places and ways (asleep, waiting, watching, playing, fighting, and so on).

Biological Thinking and the Democratic Value of Life

Agamben's account would help situate the literature of animalized or neurocultural selfhood if it could highlight the strategies that literature adopts. How does the anthropological machine help explain practical redefinitions of selfhood? For Agamben, "thinking" appears surprisingly often as a cardinal concern. At first glance, the idea of thinking developed in Gladwell's, Johnson's, and Connolly's work does not seem to fit very well with Agamben's account of the anthropological machine. Their accounts displace thinking away from conscious, logical, rational, reflective registers toward "the very smallest components of our everyday lives" (*B*, 16). Thinking is presented as something mundane, dispersed, variable, fragmentary, and transient. In this respect, their accounts very much echo long-standing philosophical criticisms of the separation of soul from body in Western thought (often originally attributed to Descartes). Do they not continue the reappraisal of thinking that Nietzsche, Freud, Martin Heidegger, or Jacques Derrida undertook?

In contrast to the philosophical critiques of mind-body dualism, the accounts of neurocultural and animalized selfhood draw on the life sciences. According to Agamben, Western political life and thought have long orbited a nucleus of animality, or "bare life." The term "bare life," or *zoē*, sourced from Aristotle, takes many guises in Agamben's work of the last decade. In general, it refers to an "incorruptible fallenness"[9] or "mere living" outside politics, norms, or judgment (*HS*, 2). According to *Homo Sacer*, modern power, today in the guise of biopolitics, persistently probes bodies at the level of "mere living." It works to subsume bare life within organized forms of life (such as modes of subjectivation or selfhood) in order to constitute itself as sovereign, constituted power.

Despite their name, the life sciences provide no direct access to (bare) life. For Agamben, they hover ambivalently on the fringes of the zone of indetermination. On this fringe, forms-of-life, organized by norms and institutions, sheer away from bare life, as Agamben notes: "Biological life, which is the secularized form of naked life and which shares its unutterability and impenetrability, thus constitutes the real forms of life literally as forms of survival: biological life remains inviolate in such forms as that obscure threat that can suddenly actualize itself in violence, in extraneity, in illnesses, in accidents."[10] On the one hand, biological lives capture the living in forms (anatomy, physiology, ecology, biochemistry, and so on). On the other hand, biological life retains something of the "unutterability and impenetrability" of bare animal life. Any turn to neurophysiology or behavioral sciences finds itself enmeshed in a complex weave of forms, norms, and exceptional events (violence, extraneity, illnesses, and accidents). These scientific knowledges supply diverse resources for social and political contests over living bodies. The animalizing accounts of thinking could extend what biological life sciences start. Outside the laboratory or the clinic, they could help render "real forms-of-life" as "forms of survival." So, while responding to a stranger, an artwork, or a political event might at first seem to lie a long way from biology, when explained in terms of a primitive or older part of the brain evolved to quickly produce fight-or-flight responses, art and politics entail forms of survival. From this perspective, popular accounts of everyday life as instinct, reaction, or survival tactic capture and exclude bare life. They, too, produce the human, even as they seek to overcome the modernist separation between mind and body.

Neurochemistry and Democracy: No Value Other Than Life

What happens to the unutterability of bare life in the animalizing accounts of selfhood? Popular accounts such as Gladwell's and Johnson's, but also academic work such as Connolly's, present certain behaviors and physiological responses as politically useful and potentially democratic. For instance, Johnson writes, "Our mental modules are implicated in political issues" (*MWO*, 213). In order to understand why "our brain's faculties may create too much resistance," we need more comprehensive accounts of "self in society" (*MWO*, 214). Hence, Johnson proposes: "There is no convincing reason a comprehensive account of self in society couldn't be built by a consilient chain: neuroscientists explain how the brain's underlying electrochemical networks function; evolutionary psychologists explain how and why those networks create channels of 'prepared learning' or instinct; . . . political theorists and moral leaders explore the best ways to structure society to reconcile those patterns of group behaviour with individual needs" (*MWO*, 214). At the base of the envisaged "comprehensive account of self in society" lies neurochemistry and, just above, evolutionary accounts of "prepared learning." In identifying and delineating the implication of brains in politics, all three authors bring certain aspects of self to the fore. Many of the examples of "rapid cognition" or snap decisions Gladwell uses come from the politicized domains of electoral politics, law enforcement, or military strategy. When they put forward a concrete site of a neurochemical-evolutionary-political-moral self in society, they typically settle on the well-known behaviors such as the fight-or-flight response. This, as Johnson writes, is partly because "learning to be afraid turns out to have been one of the most studied behavioural patterns of the twentieth century" (*MWO*, 53). Behavioral sciences value fear responses because so much can be learned from them. That is, fear heavily dramatizes the staying alive or survival of life. The neurocultural accounts value such responses as "a kind of thinking." Johnson notes: "Once again, a lack of discrimination has a potentially adaptive value. In life-or-death situations, you never know where relevant information might lie. . . . This, too, is a kind of thinking" (*MWO*, 59). This "thinking" is not linguistic, logical, rational, conscious, or even unconscious.

Many of the examples of rapid fearful cognition come from law enforcement because researchers in the life sciences have examined such responses carefully. Now these examples can become part of what Johnson calls "self-awareness" through a scientifically inflected process of personal development. The fight-or-flight response, Connolly suggests, "allows us to explore how thinking itself can sometimes modify the microcomposition of body/brain processes, as a new pattern of thinking becomes infused into body/brain processes" (*N*, 8). How can modularized, fragmentary, and partial fear responses be democratic? The account of the neurochemical-animal self in society enhances democracy by expanding the primary locus of political agency, personhood. It can think about itself more comprehensively; it can understand feelings of fear, intimations of threat, and obscure anxieties as adaptations, as forms of survival, and above all as potentially legitimate forms of thought in their own right.

These suggestions aim to enhance truth and justice. Even if these responses can fit somehow within democratic understandings of citizenship, justice, or deliberation, *zoē*'s inherently unstable position within modern democracy troubles their success. In terms of Agamben's account, they also participate in another wider process. On the one hand, as Agamben points out, "Modern democracy presents itself from the beginning as a vindication and liberation of *zoē*, and . . . it is constantly trying to transform its own bare life into a way of life and to find, so to speak, the *bios* of *zoē*" (*HS*, 9). On the other hand, the transformation of bare life into political form is not unique to democracy. Agamben claims directly that democracy and totalitarianism converge at a "historico-philosophical level" (*HS*, 10). Both know "no value . . . other than life itself" (*HS*, 10). Furthermore, he insists that only by holding on to the idea of their "inner solidarity" can "new realities and unforeseen convergences" be sensed and understood (*HS*, 10). Hence, while the animalized accounts of thinking vindicate and liberate bare life, they also transform *zoē* into a way of life or a form-of-life. In so doing, they affirm no value other than life itself.

Techniques of Transforming *Zoē*

How do the animalizing accounts transform *zoē*? Practically, they begin by *naming*. Johnson suggests learning to name chemicals and brain regions:

"If you spend some time exploring this new world, you will end up with a set of conceptual building blocks to use when thinking about how your brain works: some of them specific chemicals, some of them localized regions, some of them broader patterns of interaction between regions or chemicals" (*MWO*, 184). Naming never simply denominates. As Judith Butler argues, "To be named by another is traumatic: it is an act that precedes my will, an act that brings me into a linguistic world in which I might then begin to exercise agency at all."[11] Naming substantiates, organizes, distributes, and lays the groundwork for regulation, in this case of a sense of self oriented by biological and behavioral knowledges. For instance, Gladwell has the reader imagine going into a psychologist's office to take a language test: make four-word sentences out of some five-word sets. The sets include words like *worried*, *old*, *Florida*, and *lonely*. Presenting ten-word samples as a test for the reader, Gladwell remarks, "After you finished that test—believe it or not—you would have walked out of my office and back down the hall more slowly than you walked in. With that test, I affected the way you behaved. . . . You thought that I was just making you take a language test. But, in fact, what I was also doing was making the big computer in your brain—your adaptive unconscious—think about the state of being old" (*B*, 53). The imagined experiment sets a scene. In this scene, things take place on a largely inaccessible level (that Gladwell somewhat problematically terms the "big computer in your brain"). According to Gladwell, the language test and its aftermath—walking more slowly—show that the adaptive unconscious "picked up some clues that we're in an environment that is really concerned about old age" (*B*, 58). In trying to transform bare life into way of life, the test brings to light the penumbra of instantaneous impressions and conclusions that accompany an explicit cognitive task.

Yet visiting a psychologist's office to take such a test is not a neutral situation. Test situations generate anxieties. The norms and techniques of the psy-disciplines institute and structure the event. The imagined test in the book embeds language-thought within modes of address, compliance, norms, and performance. Taking this test, even in the literary form offered by a book on popular science, places the experience in a specific register. As Nikolas Rose puts it, "The colloquial designations, the simple examples, the dissection of recognizable moves: all these provide a means of rendering our own experiences in social transactions into thought and making them amenable to management."[12] The features that Gladwell's imagined test seeks to bring to light, and to render recognizable for readers, rely on a preexisting familiarity. Psychological testing interpellates subjects and correlates them into norms in educational, employment, medical, therapeutic, or pastoral settings. Because subjects are accustomed to being tested, they quickly recognize the structures and forces framing their responses. The test results—walking down the hallway more slowly—rely on this prior incorporation of test instruments to do the work. The test makes something recognizable (walking down the hall more slowly as the trace of a bodily modification wrought by the adaptive unconscious) because the psy-disciplines have already formed and articulated life as *bios*. Moreover, the experience of aging itself carries much biopolitical baggage. It is not one example among others. Readers may recognize the fact that multiple, autonomous adaptive behaviors shadow conscious cognitive tasking. Yet any such recognition relies on the framing provided by the prior formations that subtly corporealizes language and pervasively, ineluctably politicizes life itself.

The Territories and Behaviors of Everydayness

The experiments, situations, tests, tricks, and naming, however, do not exhaust the *bios* of *zoē*. Beyond the techniques of behavioral self-experimentation stands a more forceful attempt to overcome any separation between human and animal. For instance, Connolly contends, "Although human culture is in fact composed of essentially *embodied* beings implicated in complex patterns of action, and although some brain nodules in the human brain network are shared with other animals, cultural theorists haunted by determinist images of nature are pressed to dismiss, ignore, or degrade the corporeal layering of language, perception, and thinking in human nature" (*N*, 62). This formulation, echoed in other recent academic and nonacademic work in cultural theory and popular science, begins with commonalities between animal and human. In another example that joins animal and human, Johnson suggests, "When we sense emotional complexity in other mammals, we're detecting the existence of the limbic system operating in their brains" (*MWO*, 205). Both writers emphasize patterns of action and biological structures common to animals and humans.

At the juncture of human and animal, Agamben's exploration in *The Open* of the way Heidegger thought about differences between animal and human life again heads in a different direction. Rather than reinforcing or collapsing their separation, Heidegger's work represents for Agamben an attempt to take the animal-human separation to its historical limit. This attempt ultimately puts any separation in doubt. Heidegger's thought approaches the zone of indetermination without collapsing animal and human differences. If Agamben, via Heidegger, can sustain this movement, then an important dimension of contemporary selfhood can be reevaluated without collapsing all values and all separations.

Agamben tells how Heidegger departs from the basic concept of *territory* developed by the early-twentieth-century ethologist Jakob von Uexküll.[13] Heidegger took from von Uexküll the notion that animal time and space differs from human time and space. Animals live in milieus, and humans live in worlds. As Agamben says, "We too often imagine that the relations that a particular animal maintains with things in their milieu take place in the same time as those which link us to things in our human world" (*O*, 40; translation modified). An animal habitat consists of a system of marks that trigger perceptions and channel them into particular ways of moving through and marking out a space. Territorial limits are signaled by marks left by other animals or, in the case of some birds, by song. The relation between an animal and its environment consists of this interlocking between marks or signs and corresponding capacities to react to them in an environment. Animal milieus, according to von Uexküll, effectively shut animals in. The milieu consists of a selection of marks in close structural coupling with the senses and motor capacities of the animal in question. What falls through the sieve of these marks forms no part of the animal's milieu.[14]

The tight coupling between territory and behavior figures in the contemporary literature on animalizations of thinking. These accounts (and again Johnson's book is exemplary) take two things from it. First, they reiterate the connection between milieu and action-reaction. Second, they treat this connection as something to be experienced and acted on. People can become more sensitive to or conscious of the connection between trigger and reaction. They can also, perhaps, modify the connection. Fear, as mentioned above, and love-attachment commonly appear as the most important sites of intervention examined in these accounts. Johnson's book describes a life-threatening event that happened to the author and his partner in their Manhattan apartment. In his story of the shattering of a large window during a storm, Johnson contends that fear and his memory of that response link certain marks (the sound of wind) to reactions: "This is the body's fear response, an orchestral mix of physiological instruments launching with masterful speed and precision. . . . Feeling it kick in is one of the best ways to experience your brain and body as an autonomous system, operating independently of your conscious will" (*MWO*, 49). For Johnson, the sound of wind triggers bodily changes—tensing of muscles, sweating, shivering, eye movements, and so on.

Importantly, this trigger has also become a way for him to experience his own "brain and body as an autonomous system." Similarly, for Connolly, fear reactions offer a particularly quotidian site for the modification of thinking: "In this instance the relatively slow, complex process of perception gives way to the lightning-fast, crude processing of the amygdala.... Let's call the emergency percept *infra-perception*, because of its speed and its processing of information without visual imagery. It happens all the time, as when you turn your car in a flash on the road or freeze suddenly while walking in the woods" (*N*, 27). Perception, feeling (of fear), and reaction happen constantly, even during sleep. They are always available. However, they work without visual imagery and beneath the threshold of perception. It is hard to become conscious of them except in unusual situations, when something actually frightening occurs. The interlocking of milieu and reaction or behavior means that this "kind of thinking" has its own specificity. A sound or a flash of movement triggers a cry and a lurch sideways in response. The transition from trigger to response outstrips conscious perception or intended action. Reactions triggered by specific sensed features of a situation sweep over "intention" and "perception." Because it does not rely on language, fully formed images, or sound-images, it is fast. How, then, can the technologies of self-thinking bring it to bay?

From Excitement to Boredom

Agamben's reading of Heidegger suggests that bringing this "kind of thinking" into awareness is fraught. Because animals allegedly unify perception, action, and milieu, Heidegger famously argued that animals are "poor-in-world."[15] They do not act—they behave. Borrowing one of von Uexküll's examples, Heidegger describes how milieus lock animals in. An experimenter cuts off the abdomen of a bee and then puts the bee on the edge of a bowl of honey. The bee observes neither the superabundance of honey nor the absence of its abdomen. It just sips the honey (*O*, 52). A circuit of pulsional drives prevents the bee from seeing the catastrophe or responding to what happened. The world, Heidegger suggests, can only absorb or daze animals. Animals cannot relate to the world or things as such. Being dazed or captivated, as Agamben notes, is the basic state, the fundamental mood, of animals. This state flows from functional coupling between perceptions, a milieu understood as a system of marks or triggers that drive behaviors. It differs from human existence. Although we can be absorbed in things, we don't have to be. We act, and we have and make worlds, the argument runs, because we don't have to, because we can not do. Not having to act or do, not having to make, in short, being essentially indeterminate, grounds radical contingency, being-open, and having a world.

The amputee bee's state of absorption resonates in the recreational neuroscience texts. Experiences of love and attachment epitomize absorption for Johnson. In his book, he contrasts the fight-or-flight response to an equally or perhaps more important alternative: tend and befriend. Again, an animal model is crucial: "The prairie vole, a small rodent indigenous to the midwestern plains of the United States, is one of the natural world's great romantics. After mating, most voles remain monogamously attached to their partner for life, raising children together in a rodent version of domestic bliss" (*MWO*, 111). The prairie vole, subject of intensive laboratory investigation, teaches us about the neurochemistry of absorption, attachment, and feeling. The prairie vole's brain, in particular, turns out to be particularly instructive. "Domestic bliss" hinges on a single molecule, oxytocin: "For most people, I suspect, the neuroscience of personal connection will have more intimate revelations as we come to understand and recognize the chemicals that trigger these powerful feelings. Not just because it's intellectually interesting to know that your feelings of attachment are partially instigated by oxytocin, but also because the chemistry's effects go beyond the primary emotion itself—altering your memory, your immediate attention, your evaluation of people and environments" (*MWO*, 130). The "intimate revelations" in his account tell of his partner nursing her newborn baby in a downtown Manhattan apartment on September 11, 2001. Her calmness represents for Johnson an existential counterweight to the nervy fight-or-flight response. Oxytocin lends durability to feeling: "In other words, it's possible that oxytocin does not create the visceral pleasure of love and attachment, but it does enable that pleasure to last longer than it normally would" (*MWO*, 132). Like knowing about the amygdala's rudimentary decision-making, knowledge of oxytocin affects how we inhabit our worlds—"your memory, your immediate attention, your evaluation of people and environments." Knowing that chemicals such as oxytocin coincide with feelings of attachment affirms animality at species and individual levels: "Knowing something about your brain chemistry at such a moment connects you both to the individual neuronal assemblage in your brain that creates the image of your child and to the evolutionary history of feeling, the history of all your ancestors and their parental emotions" (*MWO*, 211). The state of absorption or captivation that Johnson attributes to oxytocin resembles the absorption that Agamben finds of interest in Heidegger.

Agamben does not equate animal absorption with the human history of feeling. Instead, through the state of captivation or dazedness (*Benommenheit*), he attempts to envisage what would happen if the anthropological machining of human-animal differences halted. Absorption exposes and expels animals into the world. Although animals only have habitats or territories, they are also, as Heidegger puts it, "expulsed" (pulsionally expelled) toward something other than themselves (*O*, 61). Their very being exposes or expels animals outside themselves. They still somewhat lack a world in this exposure, yet they engage with what absorbs them—the marks, the triggers and stimuli, that which locks in their perceptions and behaviors. In its milieu or habitat, "the animal is outside of being" (*O*, 91). This exposure or expulsion holds for Agamben an extraordinary potential: it opens the possibility of moving outside the historical impasse of anthropologization.

Agamben quickly rules out one interpretation of what this might mean: "To render inoperative the machine that governs our conception of man will therefore mean no longer to seek new—more effective or more authentic—articulations" of human-animal composites (*O*, 92). Most of the materials of neurocultural-animal selfhood risk becoming efficacious. Instead, he suggests, it would be better "to show the central emptiness, the hiatus that—within man—separates man and animal, and to risk ourselves in this emptiness: the suspension of the suspension, Shabbat of both animal and man" (*O*, 92). The weekend state of animal dazedness both sedates and animates.

Like the lasting feelings Johnson describes, the fundamental mood of animals—absorbed or dazed—is much closer to boredom than excitement. For Heidegger, profound boredom occurs when the world refuses to open itself to a constitutively open being (*Dasein*, a person). Agamben suggests that Heidegger's reading of the existentially profound state of boredom cannot ever be fully distinguished from the dazed state of animality. Conceptually and ontologically, boredom actually coalesces with the state of being dazed (*O*, 65). As Agamben concludes, "Dasein is simply an animal who has learnt to be bored, who has been woken from its own dazedness and to its own dazedness" (*O*, 70; translation modified). The mystery of the separation between animal and human lives does not rest on any essential difference such as language, rationality, capacity for invention, and so on. It lies in a zone of half-awake indetermination, between being dazed and being bored. The "banal, quotidian mysticism" he affirms comes from being dazed.[16] Effectively, Agamben, via Heidegger, folds the line between animal and human within the human in a way that emphasizes its mutability. Humans plunge into this dazed world-opening animal absorption but veer away from it frequently.

Waking and Thinking as Form-of-Life

It often seems that we must preserve thinking itself as part of human selfhood at all costs. Yet, if the technologies of animalized selfhood have any purchase, thinking cannot remain what it was. It works beneath "direct reflective regulation," as Connolly puts it (*N*, 112). Nerve-racking, stressful, or traumatic scenes—police shootouts, severe weather events, brain scans, attacks, art auctions—abound in these accounts. Thinking occurs in wild or violent events. Does the folding-inward human-animal separation in the zone of indetermination suggest any other trajectory? At the end of his discussion of Heidegger, Agamben poses two possibilities: "At this point, two scenarios are possible from Heidegger's perspective: (a) posthistorical man no longer preserves his own animality as undisclosable, but rather seeks to take it on and govern it by means of technology; (b) man, the shepherd of being, appropriates his own concealedness, his own animality, which neither remains hidden nor is made an object of mastery, but is thought as such, as pure abandonment" (*O*, 80). On the one hand, in alternative (a), the biopolitical containment of life as bare life needs to manage animality. Examples of that management litter the biomedical sciences and psy-disciplines. Johnson, Gladwell, and perhaps also Connolly take on animality through behavioral and neurotechnical practices of various kinds (MRI brain scans, biofeedback, drug and dietary regimens, and so on). Alternative (b), on the other hand, grapples with thinking itself as pure abandonment. Thinking comes close to the dazed state that lies at the heart of animality.

Animality, for Agamben, has to be "thought as such, as pure abandonment" in order to avoid either hiding or managing it. Thinking abandonment is not abandoning thinking. What would thinking be if not a startled bark into wakefulness that happens now and then? Would it be placid attachment? At the end of *Homo Sacer*, Agamben asks what to do politically, ethically, and ontologically about "bare life" and answers: "This biopolitical body that is bare life must itself instead be transformed into the site for the constitution and installation of a form of life that is wholly exhausted in bare life and a *bios* that is only its own *zoē*. Here attention will also have to be given to the analogies between politics and the epochal situation of metaphysics" (*HS*, 188). Why this injunction? Why would anyone especially want a form of life that is "exhausted in bare life," especially given that bare life is the included other that has become the principle on which biopolitical sovereignty pivots? Why must the analogies between politics and the "epochal situation of metaphysics" receive attention?

Agamben envisages the emergence of a field of research on the fringes of the biopolitical terrain occupied by politics, philosophy, medico-biological sciences, and jurisprudence. This fringe would not cut bare life into form and matter, into *bios* and *zoē*. Here, however, thought or thinking is crucial because life and form-of-life coincide in it. By contrast, the animalizing accounts attempt to identify the forms of thought that imbue life and to subject them to regulation. In an essay titled "Form-of-Life," Agamben asserts, "Thought is form-of-life, life that cannot be separated from its form; and anywhere the intimacy of this inseparable life appears, in the materiality of corporeal processes and habitual ways of life no less than in theory, there and only there is there thought."[17] Thought that can think a form-of-life that is nothing but its own existence, "a *bios* that is only its own *zoē*," would itself be a form-of-life. The connective hyphens are important. Thinking would no longer seize a historical destiny (as in Heidegger) or radically distinguish the necessary and the contingent (as, for instance, in most Enlightenment and social constructionist accounts). Nor would thinking service the increasingly fine-grained biopolitical management of ecological, economic, and political-cultural systems. Rather, thought would live, occasionally or intermittently, as *bios* and *zoē* inseparably.

An ultraminimalist, almost sedated mood pervades Agamben's alternative. As in Johnson's, Gladwell's, and Connolly's accounts, for Agamben, thought as form-of-life inhabits quotidian life. The contingencies of everyday life have been a constant theme in cultural and social theory over the last half century for various reasons (beginning, perhaps, with Benjamin, running through situationist-influenced and then feminist thought). In

Agamben's work, everyday life undergoes a further vitalization. Its radical contingency wells up along the fault lines of indetermination, where *bios* uncontrollably coincides with its own *zoē*. In the language of his earlier work, "The contingent is not simply the non-necessary, that which can not-be, but that which, being the *thus*, being only its mode of being, is capable of the *rather*, can not not-be."[18] This zone in which something becomes "its mode of being" cannot be thought except as thought by thinking itself. By virtue of its own singularity or "thusness," thinking can never become an object or a subject. In this respect, it diverges in principle at least from the techniques advocated by Connolly, Gladwell, and Johnson. If they all seek to heighten or render salient nonintentional, nonreflective thought through exceptional situations, thought as form-of-life dismantles the privilege of exceptional, shocking, or jarring events.

We have no example of form-of-life apart from thinking. Moreover, thinking leads a quotidian, not a transcendent, life. This means that it must involve, as Agamben says, "everyday zones, a very banal, quotidian mysticism."[19] Agamben's analysis of animality seeks to afford an understanding of what this quotidian life might mean. A version of bare life as everydayness can be reevaluated if we can show how separations between forms of life and life itself, or bare life, are practically produced.

Conclusion

There are many ways to sense and make sense of "our animality" or "our biological nature." Agamben provides an extremely broad account of the formation of subjectivity and political power around life. History—Western history, at least—is a constantly refashioned separation between human and animal. Any sense of humanity more or less openly depends on the twisted topology of that separation. Biological life and life as survival appear as recent versions of the excluding inclusion of bare life as forms-of-life. As biomedical sciences and politics become increasingly entwined, we can expect increasingly intimate experimental syntheses of life and power to appear.

At a more fine-grained level, Agamben offers ways of analyzing some aspects of the emerging syntheses of scientific, political, and practical life. The main features of the analysis offered here have concerned what Johnson, Gladwell, and Connolly regard as different kinds of thinking. Thinking itself begins to function as the pivotal component in the modified version of the anthropological machine. When patterns of fear, insecurity, love, and attachment come to be seen as kinds of thinking, physiological and psychological understandings of them as survival instincts or biochemical processes have to be put together with everyday experiences. Different aspects of that putting together—naming in language, practices of testing, and so on—coalesce as technologies of a thinking self.

It is important not to treat these animalizing experiments in thought solely as objects of criticism. Because they are quotidian forms of separation, they make it possible to begin to see what is at stake in the entwining of biomedical and political life. Like Agamben's readings of Heidegger, in which the structural couplings between perception, movement, and milieu also push to the limit conceptions of what it is to have a world, the animalization of thought, in its attempts to structurally couple self-awareness and biomedical knowledge, shows why it is hard to make animality present as such. (The proximity between absorption-abandonment and boredom-wakefulness that Agamben finds in Heidegger is key to this argument.) Why can't we just accept our own animality? "Our animality" cannot be thought as such because thought itself is put in question by animality. Thinking as form-of-life, something that Agamben valorizes strongly, is not very far away from management of animality. Almost the only difference is that it does not keep scientific thinking outside the frame in the way that popular and some academic accounts do.

Agamben's own affirmation of animality as a response to the "total management of biological life" (O, 77) rests on thinking the abandonment, exposure, and dazed feeling of animality as such. Across a wide spectrum of his work, he posits thinking as form-of-life as a way to activate a politically progressive relation to contemporary biopolitical power. Of all the animal figures mentioned in *The Open*, the tick that hibernated in a Rostock laboratory for eighteen years before waking to feed again is one of the most striking (O, 70). The tick woke without knowing much about where it was or about what had happened. Hibernation and suspended animation are states, however, that involve a change in immediate relations to the environment.

Notes

1 Malcolm Gladwell, *Blink: The Power of Thinking without Thinking* (New York: Little, Brown, 2005), 16. Hereafter cited parenthetically by page number as *B*.
2 Steven Johnson, *Mind Wide Open: Why You Are What You Think* (Hammondsworth, UK: Penguin, 2004), 17. Hereafter cited parenthetically by page number as *MWO*.

3   William E. Connolly, *Neuropolitics: Thinking, Culture, Speed* (Minneapolis: University of Minnesota Press, 2002). Hereafter cited parenthetically by page number as *N*.
4   Michel Foucault, *The Essential Works of Foucault: 1954–1984*, vol. 1, *Ethics: Subjectivity and Truth*, ed. Paul Rabinow (New York: New Press, 1997).
5   Steven Pinker, *The Blank Slate: The Modern Denial of Human Nature* (New York: Viking, 2002).
6   Giorgio Agamben, *The Open: Man and Animal*, trans. Kevin Attell (Stanford, CA: Stanford University Press, 2004), 37. Hereafter cited parenthetically by page number as *O*.
7   Nigel Thrift, "Re-Inventing Invention: New Tendencies in Capitalist Commodification," *Economy and Society* 35:2 (2006): 279–306.
8   Giorgio Agamben, *Homo Sacer: Sovereign Power and Bare Life*, trans. Daniel Heller-Roazen (Stanford, CA: Stanford University Press, 1998). Hereafter cited parenthetically by page number as *HS*.
9   Giorgio Agamben, *The Coming Community*, trans. Michael Hardt (Minneapolis: University of Minnesota Press, 1993), 40.
10 Giorgio Agamben, "Form-of-Life," trans. Cesare Casarino, in *Radical Thought in Italy: A Potential Politics*, ed. Michael Hardt and Paolo Virno (Minneapolis: University of Minnesota Press, 1996), 154.
11 Judith Butler, *Excitable Speech: A Politics of the Performative* (London: Routledge, 1997), 38.
12 Nikolas S. Rose, *Governing the Soul: The Shaping of the Private Self* (London: Routledge, 1999), 254.
13 Jakob von Uexküll, *Streifzüge durch die Umwelten von Tieren und Menschen: Ein Bilderbuch Unsichtbarer Welten* (*Expeditions through Animal and Human Environments: A Picture Book of Invisible Worlds*) (Berlin: Springer, 1934).
14 This closed and stable functional unity of mark, perception, or action raises all kinds of questions and gives rise to twentieth-century life sciences such as ethology and ecology. If ethology studies how signs, territories, perceptions, and actions mesh for particular animal worlds, ecology studies how the different closed milieus overlap and intersect.
15 Martin Heidegger, *The Fundamental Concepts of Metaphysics: World, Finitude, Solitude*, trans. William McNeil and Nicholas Walker (Bloomington: Indiana University Press, 1995).
16 Giorgio Agamben, "'I am sure that you are more pessimistic than I am . . .': An Interview with *Vacarme*," trans. Jason Smith, *Rethinking Marxism* 16.2 (2004): 115–24, at 118.
17 Agamben, "Form-of-Life," 156.
18 Agamben, *The Coming Community*, 105.
19 Agamben, "'I am sure that you are more pessimistic than I am . . . ,'" 118.

# Jean-Philippe Deranty

Witnessing the Inhuman: Agamben or Merleau-Ponty

A great part of twentieth-century (continental) philosophy could be characterized as the attempt to solve the conundrum of the demise of humanism. Most of the major thinkers of the second half of the twentieth century believed that humanism had become a category that no longer holds, that it collapsed with the horrific historical experiences of a mad century. They believed that the category is not just epistemologically flawed, but, more pressingly, that it is a category that is ethically and politically tainted, to the point where it can be construed as one of the key categories to deconstruct urgently in order to avoid repeating the horrors of an all-too-recent past, to step out of the possibilities of a dangerous future. The suspicion of all these thinkers is not easy to fathom, because one could apparently argue with good logic that it is precisely the normative meaning of the human that allows one to condemn the horrendous crimes committed by human beings against other human beings as failing the test of humanity. It is not an easy task to convincingly articulate why the repeated extremes of inhumanity perpetrated in the past century, culminating in the industrial planning of the total dehumanization and destruction of millions of people, also indicts with it the normative category of humanism. The core idea justifying this rejection of the normative category in the name of historical examples is that it is precisely the belief in a special human essence, implying as it does a structural separation between the human and the inhuman that has consistently helped to justify the dehumanization of so many human beings.

Added to the problem of humanism is the difficulty that none of the thinkers who denounce the perils of separation at the heart of the human believe that it would be enough simply to abandon the category. On the contrary, they all think that the task of philosophy, after the critique and the *Destruktion* of humanism, is to propose a new kind of humanism, a paradoxical posthumanist humanism that would fully acknowledge the nonhuman, yet avoid the mistake of simply erasing the difference between the human and the nonhuman. For these thinkers the most urgent task of philosophy is thus to "witness the inhuman."

The double signification of this motto highlights the difficulty of posthuman philosophy. Witnessing the inhuman means, first of all, making sure that philosophy does not bracket the historical specificity of the past century, that it sets itself as a primary task the diagnosis of the causes and the extent of the catastrophes. The first meaning of *witnessing* is simply the denunciation of the logic of separation operating at the heart of traditional humanism. Secondly, however, this program of critical diagnosis is the negative side of a positive witnessing, the one that is well articulated by Giorgio Agamben in *Remnants of Auschwitz*: "'Human beings are human insofar as they are not human' or, more precisely, 'human beings are human insofar as they bear witness to the inhuman.'"[1] For the philosophers mentioned, it is precisely the structure of the epoch that made the witnessing of the inhuman in the human impossible, that made the other inhumanity possible, the utter violence perpetrated by humans on their fellow humans.

This essay engages two attempts at solving this double task of witnessing the inhuman. In the first part, Agamben's ethics of testimony is situated in his general project of a deconstruction of what he sees as the inherently dangerous metaphysics of Western *logos* conducive to the destructive biopolitics that found in Auschwitz its fateful realization. Rather than a critique at a conceptual level,[2] Agamben's overall project is questioned here on the basis of its practical consequences, the purely negativistic, "evanescent" ethics, politics, and aesthetics that arise out of his radical critique of metaphysics. Against such a disempowering of praxis, Maurice Merleau-Ponty's alternative mode of constructing the witnessing of the inhuman is upheld. Merleau-Ponty's critique of humanism is the most fruitful one, I claim, because it is the one that opens the most promising horizon for truly effective models of ethical, political, and aesthetic praxis.

Witnessing the Inhuman: Agamben

Witnessing the inhuman for Agamben has the two intimately related meanings already mentioned: the critical task of bearing witness to the reality of the time, in all its past and potential horror, and the descriptive task of identifying the inhuman in the human, with the idea that the fateful expulsion of the inhuman from the human has been the origin of those sociopolitical frameworks that produced the inhuman politics of the time.

To the first task, an entire book is famously dedicated, *Remnants of Auschwitz*. The book has the ambition, as Agamben writes in the introduction, of offering the "Ethics more Auschwitz demonstrata," which no other book of ethics has managed to provide so far (*RA*, 13). Agamben does not see in this a pretentious or an overly ambitious claim. For him, there are compelling reasons why no book of ethics has been able so far to grapple with the historical experience summarized in the name of Auschwitz, why no genuine philosophical testimony of testimonies has yet been offered.[3] This is first of all because all other ethical principles—for example, dignity, respect and self-respect, the reciprocity of communication, Nietzschean *amor fati*—become completely irrelevant in the face of the extreme situation of the *Muselmänner*, the "drowned" human beings, as Primo Levi famously called them. Since common ethical principles make no sense in the face of the *Muselmann*, they are simply not adequate for the task: "If there is a zone of the human in which these concepts make no sense, then they are not genuine ethical concepts, for no ethics can claim to exclude a part of humanity, no matter how unpleasant or difficult that humanity is to see" (*RA*, 63–64). In fact, before the *Muselmann*, it is normativity itself, the bearer of all other models of ethics, that becomes irrelevant, thus revealing the inadequacy of all ethics that continue to operate as though Auschwitz had not happened.

There is a second, deeper reason explaining the lack of relevance of ethics to date when confronted with the extreme situation of the "drowned." This is the difficulty of testimony itself. The difficulty of second-order testimony, the one encountered by ethical thinkers contemplating the testimony of life in the extermination camps, seems to stem from the difficulty of the first testimony (RA, 33–34). But soon the specific analysis of the improbability of normative ethics in the context of Auschwitz rejoins a deeper strand in Agamben's thought, one that demonstrates the "impossibility of bearing witness" (RA, chapter 3) in general and eventually is synonymous with his definition of subjectivity. It is the very structure of all true testimony that it is simultaneously an absolute command, the true ethical imperative, and also essentially impossible. Auschwitz thus brings to light in the face of the most horrible of human experiments the structural impossibility of bearing witness in general, not just of bearing witness to the horror of the camps. This startling slide between the witnessing of a special event and the alleged general ontology of testimony that is also supposed to lead to a general theory of subjectivity and language, if the differences between those separate questions and two separate realms were left unjustified, could appear as an abhorrent conceptual confusion, as an irresponsible metaphysical play with an event with which it is not possible to be playful.[4]

In fact the connection between the witnessing at Auschwitz and witnessing in general is supported by a strong immanent justification within Agamben's work. *Remnants of Auschwitz* and the other volumes on *homo sacer* were informed by most of his many previous books. Many of the disturbing aspects of *Remnants of Auschwitz* dissolve when the book is relocated in this more general context. To fully grasp the startling slide in Agamben between testimony in general and Levi's testimony, one needs to return to his theories of language and the subject. It is his complex relationship with Martin Heidegger's account of metaphysics and the implications of this for language, subjectivity, historicity, and politics (to name but the most important fields) that justifies this slide.

The general movement of Agamben's thought is inspired by Heidegger first of all in the sense that its most fundamental premise is that of a history of Being consonant with the history of metaphysics. At the same time, however, Agamben aims to overcome what he considers to be Heidegger's failed attempt at stepping out of what, like him, he sees as a disastrous history ending in catastrophe. The most decisive task of thinking, for Agamben, consists therefore in a gesture characteristic of poststructuralist thinking, in deconstructing the remnants of metaphysical thinking that continue to operate within the Western world, in warning of the pernicious effects still to be expected from an uncritical use of discredited categories: "We live today at the extreme edge of metaphysics, at the point where it returns—as nihilism—back into its very own negative foundation."[5]

The originality of Agamben's position on this stems from his interpretation of the nefarious content of metaphysics, from the highly sophisticated and idiosyncratic way in which he interprets the nihilism of Western subjectivity and history. The central problem that he diagnoses in Western metaphysics concerns the inherently nihilistic nature of its definition of the subject and language, what he calls the "metaphysics of the Voice." The argument underpinning this denunciation has remained constant throughout his development from *Infancy and History* (1978), to *Language and Death* (1982) until *Remnants of Auschwitz* (1998) and the other books on *homo sacer*, right through to *The Open* (2002).

The starting point of his reflection lies with Émile Benveniste and the theory of enunciation. In his later work, the great linguist had corrected Ferdinand de Saussure's famous distinction between *parole* and *langage* by radicalizing it into the distinction between *semiotics* and *semantics*. The semiotic order designates the order of language as a system of established signs, while the semantic order designates the order of discourse where the semiotic elements become meaningful by being used in actual speech acts. The problem is the one attached to that passage to discourse: the semiotic order in itself does not contain the explanation for the possibility of semantic use, of a meaningful discourse addressed meaningfully by one locutor to another who understands it.[6] This is the idea most crucial for Agamben, so much so that it reappears, probably in a highly unexpected way, at the end of *Remnants of Auschwitz* (*RA*, 113–17).[7]

Enunciation, for linguistics, is the answer to this problem of the passage from the semiotic to the semantic, as it indicates and fixes the instance of discourse, the discursive reference point, from which all other reference systems within discourse (to objects, specific spatial and temporal orders, first-, second-, and third-person perspectives) take their bearing and subsequently enable the meaningful use of language in speech. Agamben's theory of language and subjectivity is the result of an emphasis on enunciation and its link with language and speech. The originality of Agamben's take on the question of enunciation, with profound implications in many other areas of his thinking, is that he interprets the shift from language to speech in a nonpsychological, indeed, an ontological, way. What this difference and this shift point to, according to him, is a "taking place" of discourse, an event of language to be distinguished from discourse itself, as meaningful content (*RA*, 114). The actual, "existential" (Roman Jakobson) reference point from which language is used in actuality does not designate an individual standpoint, a specified, human point of reference, but a "structural" dimension of language, that is to say, a dimension of language irreducible to the order of meaning and grammar, a dimension prior to it. This is the simple yet necessary moment in which language shows itself to "take place," language "indicating" itself as a meaningful event prior to any consideration of the content and meaning of discourse. Agamben uses several expressions to help us understand this difference between language and speech, between semiotics and semantics, which he interprets in terms of a structure of language, by characterizing it as a difference between saying, the event of language, and the said, the content of discourse, or as the difference between indication and signification, or between showing and saying (*IH*, 46).

In *Infancy and History*, the implication of the theory of enunciation for the theory of subjectivity is phrased in terms of "infancy." From Benveniste, Agamben borrows the thesis according to which subjectivity—the capacity for the subject to refer to itself in the first person, to take the "ego" position—is possible only through the use of a potentiality provided by language rather than on the basis of reflective experience. The *I* is whoever *says* "I" (*IH*, 49). The *I* is thus the paradigmatic example of the system of enunciation: a moment when discourse takes its place, sets itself up, as it were, as the condition for all ulterior meaningful speech. The implication of these distinctions for the theory of subjectivity in its relation to experience is drastic, as it means that there is an irreducible gap between the human and the linguistic. If accessing the order of language entails a moment of enunciation that is only the system of signs setting itself up in a self-referential circuit (the pronouns and deictics make sense only as pointers within the overall system of language; they have no objective reference external to this system), then the whole domain of experience disappears in enunciation. This is precisely what *in-fancy* means: that state of being human outside of language use. It is important to note that it is prior to language use and not to language in general, as enunciation is only what makes the passage from language to speech possible. In other words, human beings only ever start to speak on the basis of repressing their own experience. On the one hand, "It is through language that the individual as known to us is constituted as an individual" (*IH*, 49). Yet on the other hand:

> If the subject is merely the enunciator, . . . we shall never attain in the subject the original status of experience. . . . On the contrary, the constitution of the subject in and through language is precisely the expropriation of this "wordless" experience; from the outset, it is always "speech." A primary experience, far from being subjective, could only be what in human beings comes before the subject—that is, before language; a "wordless" experience in the literal sense of the term, a human infancy, whose boundary would be marked by language. (IH, 47)

This startling distinction between the human being and the subject, between the human and the linguistic, is construed by Agamben through a logic that is characteristic of his thought, one that anticipates the famous later analysis of exception as including exclusion/excluding inclusion. The subject itself is said to be born out of a suppression that maintains: the subject results from the suppression of an experience that it keeps at its heart as its very condition of possibility. Infancy is not a genetic stage preceding subjectivity—it is the "transcendental-historical" condition of it. Agamben notes: "The experience, the infancy at issue here, cannot merely be something which chronologically precedes language and which, at a certain point, ceases to exist in order to spill into speech. It is not a paradise which, at a certain moment, we leave for ever in order to speak; rather, it coexists in its origins with language—indeed, is itself constituted in the very movement of language which expels it in each instance to produce the individual as subject" (IH, 48; translation modified). As Agamben makes clear, speech and infancy are linked in the most substantial way: as the expulsion of infancy, speech relies precisely on the latter in order to be successful, as the negative that it requires even as it represses it. "Infancy is the origin of language," infancy is language's transcendental limit, but on the other hand, *language is the origin of infancy* (IH, 48), as the event of expulsion that constitutes what it separates through its constitutive boundaries. This is isonomic with the logic of the creation of bare life by sovereign power: each is the origin of the other.

Given that the subject as a whole is seen as the result of the acquisition of deictic and more specifically pronominal expressions, the paradoxical unity between speech and infancy can easily be generalized into an overall anthropological argument. On that account, *homo* becomes truly human as a being endowed with language, only at the cost of repressing the very experience upon which the use of language was predicated. This, then, becomes the specific difference of the human being. Whereas animals in fact "are already inside language," "man instead, by having an infancy, by preceding speech, splits this single language and, in order to speak, has to constitute himself as the subject of language—he has to say I. Thus, if language is truly man's nature . . . , then man's nature is split at its source, for infancy brings its discontinuity and the difference between language and speech" (*IH*, 52). This discontinuity between language and speech is seen not only as the ground of subjectivity, but, in a further generalization, as the point of entry into the symbolic. Agamben holds what is in the end a rather classical philosophical anthropology whereby the theory of the access to language provides at the same time the answer to the puzzle of the symbolic function (*IH*, 59–60).[8] The shift from nature to culture itself is grounded in the split within language. From that point, a further generalization is made possible, as the theory of language leads (as in Heidegger) to a theory about the ground of historicity: "It is infancy which first opens the space of history" (*IH*, 53).

The step that leads to a negativistic theory of human history is now easy to anticipate. It reminds one of the arguments in the *Dialectic of Enlightenment*.[9] If the process of hominization and the ontological ground of historicity are to be found in a violent separation (or rather, suspension, "exception"), then historicity and historical development are vitiated from the outset. The erasure of experience in modernity is only the realization of the structural repression of the infant within the adult, which is at the heart of all symbolic use.[10]

The ethical and political aspects of Agamben's thought flow directly from these deep anthropological/ontological meditations. They locate the cause of the catastrophe of modernity at the deepest root of subjectivity itself: in a hidden chasm between the human being and the subject, between language and discourse. Only a human praxis that will retrieve the infant in the subject will avoid the fateful structural foreshortening of human experience, which is the origin of, and is brought to its terrible completion by, Western modernity. Given the decisive, indeed, exclusive, role played by language in Agamben's theory of subjectivation and hominization, the privileged form of such a praxis will be an alternative experience of language, an "*experimentum linguae*" as the preface to the French edition of *Infancy and History* puts it: "what one experiences in this *experimentum linguae* is not simply an impossibility of saying," since speech is predicated on the suspension of experience, but "rather an impossibility of speaking on the basis of language; via this infancy which rests in the chasm between language and speech, this is the experience of the very faculty of speaking, or of the power of speech itself,"[11] before or beyond actual speech. This is the experience of language as opening of Being: "It is the unhiddenness without presupposition which humans always already inhabit and where, by speaking, they breathe and move."[12] Already in the writings of the late 1970s, Walter Benjamin's theory of messianic revolution, premised as it was on an idea of "pure language" through which human language would reconnect with the language of things, would open authentically onto Being, formed the horizon of Agamben's political writings.

Published four years after *Infancy and History*, *Language and Death* works with very similar ideas but also introduces the three fundamental notions of the Voice, death, and the animal, and thus completes the diagnosis of the nihilistic metaphysics of *logos* at the background of the well-known analyses conducted in the different volumes of *Homo Sacer* and in Agamben's recent book *The Open*.

With the theme of the Voice, Agamben fleshes out the intuition at the basis of his messianic ethics and politics: the idea of the "taking place of language" that is different from, indeed, that is the condition of possibility, if a repressed one, of signifying language. As we saw, the great linguists had identified the necessity, for indication to be at all possible, that there be a "contemporaneity" (Benveniste), an "existential relation" (Jakobson) between the indicators (of time, of presence, etc.) and the "instance of discourse" expressing the message. Agamben's key innovation is to argue that this existential contemporaneity can be thought properly only in terms of a Voice as the locus in which the "event of language" takes place (*LD*, 68).

Agamben differentiates between the vocal aspect of speech in its transformation of language, from the Voice, as that aspect of language in which language truly shows itself as taking place, as the pure intention of signifying *before* the text of signification. This Voice is the medium of the "taking place" of language, the dimension of language that exceeds animal articulation (*phone*) but precedes signification (*logos*) (*LD*, 72). With this, Agamben simply reiterates the crucial distinction noted between *significability*, the potential to signify, and *signification*. He once again isolates a pure potentiality of signification as a dimension not always already encompassed in the signification itself. This opening of significability as separate from signification has extremely important resonances in his entire work. Since, following Heidegger, he sees the opening of Being in the experience of language, the experience of pure significability beyond signification corresponds to ontological difference itself (*LD*, 58). But Agamben's added twist is that the new *experimentum linguae* is also an experience of the irreducibility of potentiality to actuality, an experience already characterized by Heidegger in his readings of Aristotle.[13] Agamben's negativistic narrative of Western *logos* and his search for an alternative metaphysics that would not be grounded on a nihilistic foundation is predicated upon this critical reading of potentiality. It points to an experience of language as the experience of the power of language before its reduction in speech and discourse. This in short indicates an alternative experience of the Voice as the locus of the pure event of language.

Agamben, however, does not spare Heidegger himself, despite his strong reliance on him. The Voice, as the "category of categories" (*LD*, 73), is the name for the clearing of Being itself, but according to him, it has only ever been analyzed in terms of negativity by the Western tradition, including and most especially by Heidegger. It is the fact that, through an interpretation of the Voice as the voice of death, negativity has been made constitutive of Being itself, which ultimately is responsible for the nihilism at the heart of metaphysics.

This negative core of metaphysics is to be found first of all in the double negativity structuring the Voice, as an instance in between the no-longer-being animal, the Voice as *phone*, and its not-yet-being signifying language, the Voice as *logos*. Agamben focuses especially on the two grand attempts at devising a postmetaphysical metaphysics, Hegel's logic and Heidegger's ontology. In both cases, a constitutive negativity operates at the heart of human subjectivity, the instance in and through which the spaces of truth and Being are to be opened. From the special place of these two grand reconstructions of metaphysics within the latter's long history, Agamben can conclude that the Voice, as the category of categories, as the locus of enunciation (of Being), has always been thought in that tradition as negativity, as the Voice of death. Given the absolutely central place of the Voice in the very process of hominization and for the passage to second nature, Agamben can conclude to the essential nihilism of Western metaphysics, to the fact that its theory of subjectivity and its characterization of the "clearing of Being" is based on a deadly foundation: "The mythologeme of the Voice is the original mythologeme of metaphysics; but inasmuch as the Voice is equally the original locus of negativity, negativity is inseparable from metaphysics" (*LD*, 162).

Against the background of these early reflections, the later texts on biopolitics take on a specific light. They elaborate in the political and the juridical the general narrative of the nihilism of Western metaphysics that was developed in studies on subjectivity, language, and historicity. At the basis of Agamben's analyses of the law, the state, and the citizen lies this fundamental premise that Western metaphysics is a metaphysics of death, a deadly metaphysics. The structure of sovereignty and the fundamental operations of the law are perfectly isonomic with the more general structure of the metaphysical construction of language, which also provides the key to the most general structure of the passage from nature to the symbolic. In all those cases, the same vicious "transcendental-historical" logic is at play, whereby symbolic institutions (signification, normativity in the figure of moral norms and political laws) rely on the repression of that which makes them possible. In all these cases, the live principle (the Voice, *bios*, significability) is included through its expulsion by the institutions (language, the state, culture). In other words, the dialectic is at the heart of Western culture and spreads its deadly negativity over it, culminating in radical death at Auschwitz.

The solutions to this deadly metaphysical logic presented in Agamben's later studies are the same as those suggested in the first two books: to step out of history since history itself is premised on repression and negativity; to step out of the logics of law and sovereignty as they operate in the same way as *logos* capturing/negating its own "infancy." Instead, salvation is to be sought in a different *experimentum linguae*, which has its prime examples in particular works of poetry that operate beyond the distinction of *signifier* and *signified*.[14] The aim is to establish a new "ethos," in the sense of Heidegger's "Letter on Humanism," a new relation to Being, one, however, in contrast with Heidegger this time, which is no longer mediated by death.[15] This is where *Remnants of Auschwitz* ends: to speak the impossibility of speaking, to find redemption in a new use of language, which will be the testimony to the impossibility of being a subject.

Against this rich and complex construct, the task of "witnessing the inhuman" has two dimensions that can be discussed separately for the sake of analysis. The inhuman is, first of all, that ground of subjectivity that is repressed as the subject becomes itself by accessing symbolic structures, foremost among them, language. Agamben calls this "experience" and "the human" in contrast with the linguistic. But if by "human" we understand the common notion of a separate creature arising from nature through its access to symbolic functions, then the prelinguistic human must be designated as inhuman. Witnessing the inhuman in this sense is the task to which *Remnants of Auschwitz* is also dedicated, beyond the testimony of radical dehumanization through the horror of the camp.

But there is a second aspect to witnessing the inhuman that *Infancy and History* and *Language and Death* have already discussed at some length. This time, the rupture that leads to subjectivity, culture, and historicity is not considered as the rupture within the human between the presubjective and the subjective, the infant and the subject. It is the rupture, within the human, between the human and the animal.

These two different ruptures correspond to the two possible meanings of anthropology: first as the study of the different ways of accessing and making use of the symbolic function, and second as the comparison between the human and the animal. The analyses in Agamben's first books are not just prolonged and developed in his studies in *Homo Sacer*. They are also pursued in *The Open*, with the deepening of the problem of the human's relationship to the animal.

The structure of negative suspension is again at play here, the logic of an inclusive exclusion, which creates a zone of indistinction where the law of exception—that is to say the possibility of total violence—is sovereign. In this case, the indistinction is that between the human and the animal: the human being becomes human by separating within itself the human from the animal and by founding the former on the repression of the latter. This is what Agamben calls the "anthropological machine": the nonhuman is produced by separating and expelling outside of the human what is supposed to be beneath, yet also somehow immanent to, it. Again, negativity is the core motor of this dialectic: the human opening of and onto the world is possible only on the basis of the negative experience of a capture and repression of animal life.

This logic has a harmless face when it functions as the production of the animal as that which exists within the human (*logos*) as the not-yet-human (life). It is the same logic, however, that continues to operate when a less-than-human is identified as a being with human traits who is not fully human. It is obvious, given the metaphysical underpinning of these distinctions, that for Agamben there is a seamless continuity between the two "productions." Given the logic of the anthropological machine, it is a mistake to think that one can escape it first by acknowledging the animality of the human and then, in a second moment, by passing on to humans' "second nature," in other words, their sublimation of animality in the production of a truly human, historical world.[16] This mode of argument fails in the same way and for the same reason as does the overcoming of metaphysics.

Instead, what is required is an exit from the anthropological logic, one that will bear the same formal traits as the new experience of language beyond enunciation and the new experience of the community beyond sovereignty. Like those other emancipatory experiences, the new anthropological experience entails a paradoxical relationship to its opposite, one that is no longer dialectical. This is another witnessing of the inhuman, no longer the transfiguration of the inhuman, but a new experience of the separation that does not end up in "suspension" or "exception." Agamben states: "To render inoperative the machine that governs our conception of man will therefore mean no longer to seek new—more effective and more authentic—articulations, but rather to show the central emptiness, the hiatus that—within man—separates man and animal, and to risk ourselves in this emptiness: the suspension of the suspension."[17]

Witnessing the Inhuman: Merleau-Ponty

To oppose Merleau-Ponty's late work to Agamben on the specific issue of witnessing the inhuman might appear arbitrary and difficult to justify. However, a number of considerations make this confrontation plausible. Merleau-Ponty also aimed to overcome what he saw as the inherent contradictions of contemporary humanism.[18] Like Agamben, his final aim was an alternative ontology that would not repeat the contradictions of traditional as well as Husserlian and Heideggerian metaphysics. Merleau-Ponty's alternative ontology involves, like Agamben's, a rethinking of the human in its relation to other beings and especially in its relation to animals. It is also premised on an original philosophy of language. Like Agamben, Merleau-Ponty saw the core issue in critiquing metaphysics and in establishing an alternative, ethically and politically more satisfying ontology in the correct characterization and location of negativity. In fact, it could be argued that this, the importance accorded to negativity, is the precise point where the two projects are most comparable, and where they diverge the most. It is also the point where Merleau-Ponty's alternative proposal uncovers the origin of the limitations in Agamben's ethics, politics, and aesthetics. Their radically differing understandings of negativity lead to two extremely contrasted ontologies: one, Agamben's, structured around transcendence, and the other, Merleau-Ponty's, built on immanence.

The first step in characterizing this opposition is to start by noting that Merleau-Ponty's fundamental methodological premise is the rejection of ontological difference as difference. What he calls in his lectures on Nature the "binocularism" and the "strabism" of all philosophy (including Edmund Husserl's and Heidegger's) is the dichotomy between the total order of Being (God, the in itself, Nature, the realm of objectivity) as opposed to the order of existence that escapes the total immanence of Being (the created world versus the principle of creation, beings and not Being, the subject and not the object, etc.): "What is given is the metamorphosis of brute being, the giving birth. We are going to Being by passing by beings. . . . There is a circular relation between Being and beings."[19]

Second, as the mention of metamorphosis and birth intimates, central to this indirect ontology is a *productive* conception of negativity. Merleau-Ponty arrives at a productive definition of negativity because he links it directly to a conception of life that goes beyond the traditional antinomy of mechanism and finalism. Finalism is discredited by modern science, even though the correct characterization of the organism verges precariously on it. Mechanism, on the other hand, which seems to be natural science's logical conclusion, also falls short of the complexity of organic life. For example, it cannot explain animal embryonic development as studied by modern biologists, who highlight the fact that there is at some point of ontogenesis a general organic integration *before* the full development of the neural system, a fact that proves that it is reductive to describe animal ontogenesis in terms of neurological causality (N, 139–67). Instead, one must conclude that the animal organism is, in fact, best explained in its functioning and genesis as the locus of behavior, in which the totality transcends the parts by integrating them into one general "theme," the idiosyncratic, "melodic" mode of being in and coming to grips with that entity's environment.

Merleau-Ponty's negativity stems from this notion of a general behavioral scheme specific to each organism, which is anticipated in the development of the organism and retroactively, but not teleologically, guides it: "We must place in the organism a principle that is negative or based on absence. We can say of the animal that each moment of its history is empty of what will follow, an emptiness which will be filled later" (N, 155). And in a later passage, he writes: "The true nothingness is not *nichtiges des Nichts* [the nothingness of the Nothing], but an *Etwas* [something] always on the horizon, the positive determinations of which are the trace and the absence" (N, 255). This definition of negativity as the absence of a meaning to come, which haunts the present and guides it already, characterizes organic life. Crucially, it has the exact same structure as expression.[20] This is Merleau-Ponty's fundamental intuition: that life is expressive and expression is life, however sophisticated or primitive the forms of life, to whatever degree of complexity this power is incarnated; in other words, life is the origin of meaning and symbolism.

As life is expression, all living organisms can be shown to be expressive in terms of their behavior. An even more radical implication of this principle of expression is that the most complex forms of organic integration ("higher animals") *share* a fundamental faculty with the lower ones. Higher organisms replicate and complicate in their interactions with the world and among themselves an expressive capacity that was already at play, however crudely it was effected, in lower organisms.

The view of organic life as opening, qua behavior, a dimension of meaning within the material world enables Merleau-Ponty to develop an immanentist philosophy of nature, where the superior levels actualize potentialities that were already inscribed in the lower ones. In the *Nature* lectures, for example, the reading of G. E. Coghill's study on the axolotl leads to the fundamental insight that the organism and the behavior are one, which makes the body no longer an inert mass of anatomy but a carrier of meaning.[21] This then leads to the reading of Jakob von Uexküll and the well-known comparisons between lower and higher animals. Here the already complex reaction of lower animals to the external world, understood no longer in the causal terms of stimulus response but as the specific theme of that animal's organic schema, leads without discontinuity to the even more complex delayed and fine-tuned response of higher animals. These, Merleau-Ponty argues, react no longer to stimuli but to signs, in such a way that without their active involvement (their movements, their behavior) those signs would not exist as such. In other words, higher animals create forms of meaning, a protosymbolism within nature. This is one way of interpreting Uexküll's famous notion of *Umwelt*. The upshot is: "At the stage of higher animals, the Umwelt is no longer a closing-off, but rather an opening. The world is possessed by the animal. The exterior world is 'distilled' by the animal, who, differentiating sensorial givens, can respond to them by fine actions, and these differentiated reactions are possible only because the nervous system amounts to a reproduction of the exterior world, as a 'reproduction' or 'copy'" (N, 171).[22] Merleau-Ponty does not hesitate to characterize the perception-reaction of higher animals as the first *opening*, or clearing, of the world, before human intervention. This is in direct contradiction to Heidegger and, following him, Agamben, for whom "the open" is the reserve of *Dasein*. More generally, Merleau-Ponty locates the origins of symbolism, culture, and communication in the instinctual behavior of higher animals (N, 195). This, obviously, is in stark contrast with Agamben's logocentric philosophical anthropology, which identifies in the acquisition of language, in enunciation, the passage to second nature.

Merleau-Ponty's ontology is thus an ontology "in spiral," which seeks to highlight the "verticality" of Being, such that no discontinuity or transcendence needs to be introduced between any of its successive "layers."[23] The vector carrying from one "layer" to the next, from nature to spirit, is life itself, in its indivisible power as expressivity, and the fuel of that vertical continuity is "negativity" understood as meaning-to-come. In this immanentism, negativity is not equated with death but as the principle of expressive life. The special "ontologies" are not only continuous but isonomic; they "concern the leaves of one sole Being which can be globally defined as what is not nothing" (N, 220). To put it in terms of the interpretation of Hegel (an essential moment in Agamben's diagnosis of metaphysical nihilism), Merleau-Ponty, contrary to Agamben, simply accepts Hegel's idea that negativity is the heart of life and meaning.

The top of the spiral is occupied by the passage from the visible to the invisible, where *logos* is thus shown to be rooted in life.[24] The privileged point of passage between the visible and the invisible is obviously the human body conceived as flesh. The flesh, as is well-known, is the body as it makes the experience of reversibility, the experience that it is both sensing and part of the world that it senses, the realization that it stands "between what is in front and what is behind me" and is thus "in circuit with the world." As a result of this, the integrated postures of the body schema are not just "subjective" adaptations to the *Umwelt*, but they "map out" the world itself, such that the human flesh becomes the indicator of the flesh of the world (N, 223). In the human organism, the feedback mechanism already at play in other higher animals is developed to its full potential and becomes a full "circuit," such that "perceived things [are] correlations of a carnal subject, rejoined to its movement and to its sensing; interspersed in its internal circuit—they are made of the same stuff as it. . . . The flesh of the body makes us understand the flesh of the world" (*N*, 218). This vitalist ontology of the *Ineinander*, where things continually step *into* each other, contradicts frontally the main features of Agamben's post-Heideggerian project premised as it is on a strict adherence to the ontological difference. The dualism of the human and the animal is no longer needed.[25] For Merleau-Ponty, there is a way of conceiving of all life as *bios* without having to resort to any definite separation between the human and the nonhuman. Since for him life is already expressive, human language and all symbolic institutions more generally are rooted in the meaningful totality of organisms that are symbolically, and even semiotically, linked to their external and social worlds, qua integrated forms of behavior. There is no need, therefore, to oppose language and experience. The rejection of such a dichotomy is the basis for all of Merleau-Ponty's writings on language, from the famous sixth chapter in the first part of *The Phenomenology of Perception* ("The Body as Expression and Speech") to his later "Saussurian" writings, most notably in "The Phenomenology of Language."[26] His philosophy of language highlights the seamless passage from the experience to language, and his later work is dedicated specifically, as is well known, to this doubling of the visible by the invisible, the "lining" of embodied experience with its symbolic reappropriations, in linguistic, artistic, scientific, and philosophical expressions.

Such a philosophy of language has its own take on the question of enunciation. It interprets the "contemporaneity" between indicators of presence and the instance of discourse as the ambiguous relationship between a lived experience that creates the fundamental "field of presence" from which discourse makes sense, and the given, instituted languages that partially dispossess the individual act of speech. Merleau-Ponty does not deny the desubjectifying effect of language use, but instead of concluding from this to a radical chasm, he emphasizes the difficulty of true expression, the difficulty of transforming an impersonal, self-relating language into a vehicle of a unique, embodied experience, a difficulty, or a dialectic that is structurally the dialectic of life itself.

In such an ontology, witnessing the inhuman becomes the task of being true to the structural empathy that allows the human flesh to be in tune with the flesh of the world. This means not just a retrieval of the structural interhuman intercorporeity at the basis of the human flesh. It designates also the *Ineinander* of the human and the animal and, even more radically, that "I am with things in a rapport of *Einfühlung*: my inside is an echo of their inside" (N, 224). To be fully human on this account is to welcome the nonhuman in one's flesh, to recognize one's full presence and participation in the sensible world.

The critical counterpoint to this understanding of the inhuman is that it is inhuman, in the normative, critical sense this time, to establish separations where there are none. Merleau-Ponty has no special theory of Auschwitz. He has no separate moral theory, no theory of evil in particular. This is all linked to his phenomenological method, which describes constitution and can provide only an implicit account of normativity.[27] This it does, however. Merleau-Ponty's powerful model of the constitutive nonhumanity within the human embodiment points indirectly to an ethics of being in the world, where the ethical duty is to be a full participant recognizing the existence and right of all that participates in one's own flesh. Critically, this would allow for a diagnosis of those historical phenomena in which individuals and communities have reneged upon their inhumanity by creating abstract separations between themselves and the world, the world of other human subjects, of animals, and of nature itself.

Politically, Merleau-Ponty's writings on historical praxis simply draw the consequence of the ambiguous status of the human being in the world, as a being that is both able to create institutions and change history (be expressive), but on the basis of a radical passivity, as a being to whom meaning and institutions are also *given* and who must constantly measure the extent of her own ignorance and powerlessness. This is his understanding of the dialectic at the heart of historical praxis, as a self-criticism that can never be interrupted but that is also conducive to meaning-creative action.[28] This produces a blueprint for a substantive philosophical justification of politics as historical praxis, where the ambiguities and uncertainties of collective and individual action—immersed as they are in the infinitely open and complex horizon of history—are fully acknowledged, yet the necessity and the promise of action are also upheld.[29]

In terms of the content of praxis, the ethical imperative of an active acknowledgement of one's full participation in the world was interpreted by Merleau-Ponty as synonymous with a critical (Marxist) humanism that would be guided by the goal of the abolition of inequality and injustice.[30] But since the active participation in the world also entails a fundamental empathy with the nonhuman, his politics easily lends itself also to a translation in a radical political ecology. Bruno Latour's argument in favor of a political representation of all beings involved in, and affected by, our actions could be seen as one possible development of Merleau-Ponty's fundamental insights.³¹

Finally, a substantive aesthetic perspective is also opened by Merleau-Ponty's vitalist ontology. This is an aspect of his later thought that is well studied, especially the great incisiveness and originality of his writings on painting, but we should not forget the importance of literature for him, especially of the novel. The ontology and ethics of full participation in the world emphasize constantly, as one of its key operating concepts, the miracle of expression. The works of art and in particular truly creative linguistic expression are used by him as paradigmatic cases, notably in *The Prose of the World*.³² The expression achieved in works of art stands out as being close to miraculous, because, as in the paradigmatic example of Paul Cézanne, the "expression of what *exists* is an infinite task," as it is at first an insoluble problem to be able to do justice through artistic expression to the expression that the world itself is.³³

The ideal of a full embrace of the world that inspires Merleau-Ponty's ontology of immanence, its resulting ethics and politics of engaged participation, and the substantive forms of praxis that they underpin contrast starkly with the evanescent negativism characterizing Agamben's ethics, politics, and aesthetics. Because, like Heidegger, Agamben locates the nihilism of modern culture in the metaphysical nature of its *logos*, his critique is nothing but radical. It approaches with uncompromising suspicion all categories, all logical links between them, all previous modes of thinking reality, of articulating experience, for their suspected collusion with death, as all converging toward Auschwitz, beyond all ontic differences that become insignificant.³⁴ Conversely, because the critique is pitched at such a radical level, a nonnihilistic approach to the world and to human experience, nonnihilistic forms of praxis that could be based on them are available only by stepping out and going into the beyond. But since the beyond understood in a straightforward sense would simply reinstate the logic of separation and dichotomy, Agamben's beyonds are paradoxical ones that simultaneously state and deny their attachment to what they leave behind. They all function on the basis of a paradoxical logic, the logic of suspension, which allows him to evoke realms that are beyond the identity or difference of identity and difference, realms that posit and deny at the same time unity and separation. Ethics becomes a practice that is not practical: it involves no action and only contingently attitudes toward others. It is a normative attitude beyond the normative, an experience of the subject's relation to itself, where the latter speaks its own impossibility. The experience of language it relies on is to testify to the metalinguistic core of language. Politics is a practice where the notion of praxis is supposed to have been made redundant, a historical experience beyond history, a communal action that rejects the notion of community, a challenging of the law beyond the notion of the law. Poetry and literature are uses of language beyond the signifier and the signified, suspicious of the deadly, hidden ground of expressivity.[35]

It is difficult to brush off the impression that, despite its amazing sophistication and erudition, Agamben's philosophy only leads to an evanescent theory of praxis that has little to say and indeed is not really interested in having anything to say to and about real practices. Merleau-Ponty's full-blooded ethics of an engagement with the world—premised as it is on an understanding of dialectic as the challenge of ambiguity, of a negativity that is the promise, possible but always uncertain, of instituting new meanings—seems to outline a much more fruitful grammar for productive practices. If witnessing the inhuman is to remain more than a philosophical exercise, it is in substantial forms of praxis that it will have to be realized.

Notes

1. Giorgio Agamben, *Remnants of Auschwitz: The Witness and the Archive*, trans. Daniel Heller-Roazen (1998; New York: Zone Books, 1999), 121. Hereafter cited parenthetically by page number as *RA*.
2. I have attempted to develop such a critique in "Agamben's Challenge to Normative Theories of Modern Rights," *Borderlands* 3.1 (2004), www.borderlandsejournal.adelaide.edu.au/vol3no1_2004/deranty_agambnschall.htm.
3. The ultimate testimony to which he wants to bear witness philosophically is obviously that of Primo Levi. See *RA*, 59.
4. This is one aspect of Jay Bernstein's sophisticated critique of Agamben's "aestheticized" approach to the Shoah. See "Bare Life, Bearing Witness: Auschwitz and the Pornography of Horror," *parallax* 10.1 (2004): 14.
5. Giorgio Agamben, *Language and Death*, trans. Karen Pinkus and Michael Hardt (1982; Minneapolis: University of Minnesota Press, 1991), 99. Hereafter cited parenthetically by page number as *LD*.
6. Giorgio Agamben, *Infancy and History: Essays on the Destruction of Experience*, trans. Liz Heron (1978; London: Verso, 1993), 99–103. Hereafter cited parenthetically by page number as *IH*.
7. See also *IH*, 53–56.
8. The penultimate section of the chapter compares this discontinuity with Claude Lévi-Strauss's theory of the myth as precisely providing the link between language and speech.
9. Theodor Adorno and Max Horkheimer, *Dialectic of Enlightenment*, trans. Edmund Jephcott (Stanford, CA: Stanford University Press, 2002).

10 There is no space to compare this theory of infancy with Jean-François Lyotard's late writings (particularly in *The Inhuman*). The overlaps are many and significant. Lyotard, *The Inhuman: Reflections on Time*, trans. Geoffrey Bennington and Rachel Bowlby (Stanford, CA: Stanford University Press, 1992).
11 Giorgio Agamben, *Enfance et Histoire: Destruction de l'Experience et Origine de l'Histoire*, trans. Yves Hersant (Paris: Payot, 2002), 14–15.
12 Agamben, *Enfance et Histoire*, 5–18.
13 This idea of a pure potentiality beyond its reduction to actuality (which generalizes the linguistic distinction between *semiotic* and *semantic*) is one of the threads running through Agamben's entire body of work. It is a key argument in *Homo Sacer: Sovereign Power and Bare Life*, trans. Daniel Heller-Roazen (Stanford, CA: Stanford University Press, 1998), 44–48. It is substantially related to Heidegger's late ethics of *Gelassenheit*, of "letting be." See Giorgio Agamben and Valeria Piazza, *L'Ombre de l'amour: Le concept de l'amour chez Heidegger* (Paris: Payot, 2003), 45.
14 Hence there is a substantial link between the political and the poetic in Agamben. See especially Agamben, *The End of the Poem: Studies in Poetics*, trans. Daniel Heller-Roazen (Stanford, CA: Stanford University Press, 1999).
15 Martin Heidegger, "Letter on Humanism," in *Martin Heidegger: Basic Writings*, ed. and trans. David Farrell Krell (New York: HarperCollins, 1993), 213–66.
16 Heidegger was for him the "last philosopher" to still believe in this solution. See Giorgio Agamben, *The Open: Man and Animal*, trans. Kevin Attell (2002; Stanford, CA: Stanford University Press, 2004), 75.
17 Agamben, *The Open*, 92.
18 See Merleau-Ponty's 1956 lectures at the Collège de France on the philosophy of nature: Maurice Merleau-Ponty, *Nature*, trans. Robert Vallier (Evanston, IL: Northwestern University Press, 2003), 136. Hereafter cited parenthetically by page number as *N*. This translation is at times seriously inaccurate and misleading. I have altered it too often to note.
19 See also this passage from *The Visible and the Invisible*: "One cannot make a direct ontology. My 'indirect' method (being in the beings) is alone conformed to being." Maurice Merleau-Ponty, *The Visible and the Invisible*, trans. Alphonso Lingis (Evanston, IL: Northwestern University Press, 1969), 179. Note that in the same passage Merleau-Ponty describes this method as "sigetic," as an exploration of the "Abyss," while the very same notion of the "sigetic" is Agamben's early tag for the nihilism of Western metaphysics (*LD*, 155).
20 As developed most substantially, for example, in Maurice Merleau-Ponty, *The Prose of the World*, ed. Claude Lefort, trans. John O'Neill (Evanston, IL: Northwestern University Press, 1973), 41. In a later passage, Merleau-Ponty sees in the power of expression the true meaning of dialectic (120).
21 Note the very different interpretation of the axolotl by Agamben in *Idea of Prose*, trans. Michael Sullivan and Sam Whitsitt (Albany: State University of New York Press, 1995). In the regressive evolution that seems to be at play in this species, Agamben finds a natural equivalent to his idea of the structural infancy at the heart of the human. Merleau-Ponty, by contrast, following the biologists, focuses on the lessons one learns about genesis from the famous lizard.
22 Merleau-Ponty does not write "réplique à" (retort or rejoinder *to*) but "réplique de," the reproduction, or "replica," *of*.

23 Merleau-Ponty, *The Visible and the Invisible*, 178.
24 "There is a Logos of the natural, aesthetic world, on which the Logos of language relies" (*N*, 212).
25 Agamben's position on this is ambiguous since, despite his denunciation of the anthropological machine, he maintains some form of dualism because of his approach to language and his linguistic theory of hominization that separate the human realm from other realms of life.
26 Maurice Merleau-Ponty, "The Phenomenology of Language," in *Phenomenology, Language, and Sociology*, ed. John O'Neill (London: Heinemann Educational, 1974), 81–95. Maurice Merleau-Ponty, *The Phenomenology of Perception*, trans. Colin Smith (London: Routledge, 2002).
27 See, however, a passage in a newspaper article in which Merleau-Ponty rejects the identification of fascism and communism and goes on to say, "this means that we have nothing in common with a Nazi and that we have the same values as a communist. A communist, one might say, has no values. He or she only has *fidelities*. We would answer that he or she might do what he or she can to achieve this, but that, thank goodness, nobody can live without breathing. He or she has values despite himself or herself." Merleau-Ponty, "L'URSS et les camps," in *Signes* (Paris: Gallimard, 1960), 433–34.
28 This is the basic thought underpinning the critical analyses in *Adventures of the Dialectic* and his version of historical materialism. Maurice Merleau-Ponty, *Adventures of the Dialectic*, trans. Joseph Bien (Evanston, IL: Northwestern University Press, 1973).
29 Merleau-Ponty's *Phenomenology of Perception*, for example, ends with a vibrant call to action.
30 See a nice one-line summary, again in the last page of Merleau-Ponty, *Phenomenology of Perception*: "Your freedom cannot be willed without leaving behind its singular relevance, and without freedom for *all*" (456).
31 Bruno Latour, *Politics of Nature*, trans. Catherine Porter (Cambridge, MA: Harvard University Press, 2004).
32 Maurice Merleau-Ponty, *The Prose of the World*, trans. John O'Neill (Evanston, IL: Northwestern University Press, 1973).
33 Maurice Merleau-Ponty, "Cézanne's Doubt," in *Sense and Non-Sense*, trans. Patricia A. Dreyfus and Herbert L. Dreyfus (Evanston, IL: Northwestern University Press, 1964), 21.
34 Robert Sinnerbrink highlights very well the conceptual and normative leveling that occurs as a result of grounding critique in a general narrative on metaphysics and in ontological difference. See Sinnerbrink, "From *Machenschaft* to Biopolitics: A Genealogical Critique of Biopower," in *Critical Horizons* 6 (2005): 239–67.
35 See Agamben, *The End of the Poem*, 70–72.

Krzysztof Ziarek

After Humanism: Agamben and Heidegger

Giorgio Agamben's *The Open* offers a rethinking of the relation between humanity and animality with the aim of freeing them from the complex history of the conceptualizations of the human-animal relation within the discourse(s) of humanism or from what Agamben calls the "anthropological machine of humanism." Within this anthropological machine, the human is elaborated both *in* relation to the animal and *as* this very relation; that is, the human comes to be understood within the horizon of this constitutive difference from animality. Crucially, this difference becomes inscribed in the very notion of the human, so that the human comes to be conceived as essentially a human animal (*homo animalis*) or as a rational animal (*animal rationale*). After discussing various historical figures of "humanist" humanity, Agamben argues for suspending or "unworking" humanism, in particular its constant belaboring of and repeated attempts to master the constitutive divide between humanity and animality. To refigure this split in a manner that would render humanism inoperative, Agamben frames his book within a reading of a miniature from a thirteenth-century Hebrew Bible in the Ambrosian Library, which, portraying the righteous as human figures with animal heads, shows a new form of relations between animals and humans.[1] Describing this miniature as illustrating the reconciliation of humans with their animal nature, Agamben explains this reconciliation as a hiatus within "man," which comes to separate and suspend the human and the animal. "Is it not possible, therefore, that in attributing an animal head to the remnants of Israel," he writes, "the artist of the manuscript in the Ambrosian intended to suggest that on the last day, the relations between animals and men will take on a new form, and that man himself will be reconciled with his animal nature" (3). This reconciliation would effectively amount to a suspension of the need to understand the relation between human and animal and thus to a letting-be in the human of the irreducible difference between animality and humanity, a difference that cannot be brought within the purview of knowledge. "Humanity" comes to be reconciled not simply with animality and the animal within the human but also with the hiatus and the emptiness, that is, with the cessation of labor, which comes to characterize humanity after humanism. This reconciliation thus pertains not only to (the) animality (within the human) but, and perhaps even more important, to the "human" need to know and understand itself, which, to the extent that a human is a human animal, requires a comprehension and an articulation of the human-animal relation.

Guided by this figure of the human-animal reconciliation, Agamben proposes to render the humanist/anthropological machine inoperative, which would suspend the constant elaboration of humanity in relation to animality and free both the human and the animal from their entwinement: "To render inoperative the machine that governs our conception of man will therefore mean no longer to seek new—more effective or more authentic—articulations, but rather to show the central emptiness, the hiatus that—within man—separates man and animal, and to risk ourselves in this emptiness: the suspension of the suspension, Shabbat of both animal and man" (92). What comes with the suspension of the exhausted and idling anthropological machine at the end of humanism is the possibility of ceasing to repeatedly elaborate and rework humanity through and apart from animality, in short, the possibility of a nonworking of the relation that has come to structure various humanist discourses. The human after humanism is marked, therefore, by the emptiness of the caesura between the human and the animal, in which both humanity and animality are let be in what Agamben calls the "zone of nonknowledge."

This "unworking" of the anthropological machine of humanism is presented by Agamben in large measure as a critical rereading of Martin Heidegger's reflections on humanity and animality, in particular of Heidegger's remarks in *The Fundamental Concepts of Metaphysics* and his reflections on the animal and the human ways of relating to the open. Indeed, placed side by side, Agamben's *The Open* and Heidegger's "Letter on Humanism" might read like two versions of the critical question about the aftermath of humanism. For Agamben, the answer lies in rendering inoperative the anthropological machine of humanism and the resulting liberation of the human-animal relation into the figure of nonknowledge. For Heidegger, the questioning pivots on the issue of the human in relation to *Da-sein*, in particular to the *Da*, the there, taken as the site of the event (*das Ereignis*). Especially in the second part of *The Open*, Agamben presents his argument as a rereading of Heidegger's discussion of the relation between *Dasein* and animality, a reading that Agamben appears to carry out to a large extent in the "spirit" of Heidegger's deliberations. Still, in the process of rethinking the human and animal relations to the open, Agamben does not address—deliberately?—what without doubt is the crucial and most radical aspect of *Da-sein* in Heidegger's reading, namely, the fact that *Dasein* is not human but refers to an opening onto being, to a site or a place given to humans to keep open, so that being could unfold—*in* and *as* the *Da*—in the historico-temporal complexity of its event. Heidegger indeed proceeds in a different direction than Agamben, as he tries to put aside not only the human-animal relation, but also, and in a very specific way which I will explain later, the human. Agamben, by contrast, concerns himself with rendering inoperative the anthropological machine of humanism in order to save humanity from its own desire to redeem itself through grasping its relation to animality. It is not a question here of simply tracing differences between Agamben and Heidegger or of pointing out what might be either inexactness or more likely an interpretive twist in Agamben's reading of Heidegger. Rather, what is at issue turns out to be a pivotal difference in reading the "human" after humanism.

This difference becomes so decisive because Agamben's reflections on humanism, in spite of his unworking of the anthropological machine, remain marked by a certain trace of humanism, namely, by the vestige of the deeply ingrained metaphysical presumption about the necessarily central role that "human animality" plays in the understanding of the human, a presumption that *The Open* does not appear to question as such. At issue here is the priority given to comprehending the human being, first and foremost in terms of its relation to animality, that is, as a human animal, which is defined most often in terms of rationality and conceived, therefore, as essentially a rational animal. In the historical-conceptual developments of humanism that Agamben discusses, the human-animal duality is inscribed within the human and worked in ways that attempt to elucidate and master it through either comprehension or separation, or even exclusion, of animality. Whether the human is considered, thanks to *logos*, as essentially separate from the animal or as an addition to or a development (and betterment) of the animal, the human gets to be comprehended and experienced in relation to animality. This framing of the human within the relation between animality and humanity constitutes the originary gesture of humanism, which sets the scope, the limits, and the categories within which the human can be at all comprehensible. This is precisely the reason why at the end of metaphysics and humanism, the anthropological machine needs to be unworked in such a way that the human-animal relation ceases to be belabored in an effort to elucidate and master it, and is instead let be. This hiatus or "Shabbat," as Agamben calls it—essentially a suspension and a nonworking of the human-animal duality—introduces the possibility of letting go of the very idea of redeeming either the human or the animal, or both: the illustration of the righteous with animal heads portrays a poignant disjunction between the human and the animal, a disjunction, however, that no longer aims to "save" the human but, on the contrary, indicates "the inactivity [*inoperosità*] and *desœuvrement* of the human and of the animal as the supreme and unsavable figure of life" (87). Yet this critical gesture of letting the human-animal be retains the trace of the priority of this relation, albeit only as a nonworking relation. As a result, what is paradoxically left "working," as it were, when the anthropological machine is no longer operative, is precisely this *nonworking* human-animal bond. Despite the *désoeuvrement* of humanism, this inoperative bond continues surreptitiously to mark the trace of the "humanist" presupposition of the human as constituted through the—this time nonworking—relation to the animal. In this vestigial way, Agamben's reflection remains metaphysical, since it retains the optics of humanity's "primary" relation to animality by presenting, as it were, its verso, where the hectic working and the desired mastering of this relation is suspended and given a break, a Shabbat. Differently put, humanity in *The Open* remains vectored toward animality, as is clearly underscored by the illustration from the Bible, which seems not only to prompt but in the end also to circumscribe Agamben's brilliant and erudite analytic unworking of the humanist-anthropological machine. For rendering the human-animal bond inoperative does not in fact sideline or sidestep it. As a matter of fact, if we are to take the biblical illustration at its "figural" value, it reinforces this bond, giving it more power and binding force, precisely because humanity as well as animality come to be "saved" by leaving the human-animal relation nonworking, releasing it into the zone of nonknowledge.

The portrayal of redeemed humanity in the biblical miniature presents in fact a metaphysical and, to that extent, an anthropological vision of the human: while the animal "part" is finally freed from the human need of mastery, thereby also freeing the human from the same need, the miniature reconfirms the human as the "inoperative" human-animal. Agamben finds himself here at the threshold of letting go of the double—human and animal—determinations of the human being, and in this specific sense, the end of *The Open* can be seen as leading in a direction parallel to Heidegger's questioning. Looked at more closely, the hiatus delineated by Agamben suspends the reciprocal elaboration of the human and the animal, not only leaving them in the zone of nonknowledge beyond knowledge and ignorance but in fact turning them into the very figure of such nonknowledge.

"Nonknowledge" remains a difficult and enigmatic notion throughout *The Open*, and Agamben describes it differently in various contexts. One way to think about this concept is as an alternative to the binary opposition of knowledge and ignorance. While not a form of knowing or understanding, nonknowledge is also not ignorance or lack of knowledge. Rather, it indicates that the human-animal relation lies beyond the question of knowledge and its absence, or to put it differently, that it is not of the order of knowing. This is the case because the human-animal relation is of the ethical order for Agamben: it is something to be experienced or lived ethically rather than submitted to intellectual inquiry and the parameters of knowledge. One can also approach nonknowledge another way: knowledge and/or ignorance are the par excellence human(ist) parameters for elaborating and understanding the human relation to (its own) animality. As such it encloses both the human and the animal within the humanist optics of understanding, without leaving any other option for figuring this relation. Within humanism, the human-animal relation must be determined as either something that is known and understood, or something that, though still unknown, can become known, that is, something that can be worked out in terms of human(ist) understanding. Nonknowledge indicates the interruption in this humanist approach to the human-animal relation, the caesura that puts knowledge and ignorance aside, and thus constitutes an opening to another mode of being human-animal, one not determined by or concerned with understanding from the start. Nonknowledge becomes an alternate framework for the human-animal relation, one in which this relation ceases to be an issue, certainly an issue *for* and *of* knowledge.

No doubt both the human and the animal become profoundly altered by this displacement beyond the humanist optics of knowledge. In fact they are altered beyond knowledge and ignorance, as Agamben makes clear (90–91), so that they can no longer be referred to as simply human or animal, since these terms are the cornerstones of the humanist conception of the human: "no longer human or animal contours of a new creation" (92). As Agamben concludes, "The righteous with animal heads in the miniature in the Ambrosian do not represent a new declension of the man-animal relation so much as a figure of the 'great ignorance' which lets both of them be outside of being, saved precisely in their being unsavable" (92). Agamben thus succeeds in releasing the human-animal relation from the hold of knowledge (and of its obverse side, ignorance) and, in the same gesture, saves the human-animal from the purportedly redemptive operations of the anthropological machine of humanism. It is the rendering inoperative of the anthropological machine and of its "redemptive" practices of mastery that, in fact, saves the human-animal and saves it precisely as unsavable, that is, as outside of the *logos* of salvation. If it is the optics of knowledge that articulates the "human" as human and thus as a human animal, then within the zone of nonknowledge disclosed by Agamben, the "human" can no longer be "simply" human, in the specific sense that it no longer submits its "essence" to its articulations, comprehension, and knowledge in terms of "humanity." The same would obviously pertain to the animal, no longer understandable in terms of animality but instead "aknown" and, by extension, to the human-animal relation determinative of humanism.

This transformation, however, retains these altered versions of humanity and animality as the crucial dimensions literally shaping the "saved" humanity, as the illustration from the Hebrew Bible indicates: saved precisely by being left out of the workings and of the *logos*/logic of salvation, understood here as being brought into light and comprehension. Clearly, the suspension of the anthropological machine of humanism means that one no longer seeks new articulations of the human, which, produced as successive elaborations of the difference between the human and the animal, become themselves inscribed in the existence of human beings. What is of decisive importance here, however, is that even though they have been rendered unworkable, the relation and the difference between the human and the animal remain formative of the no longer simply human, or animal, humanity illustrated by the miniature. It is true that the human-animal relation ceases to shape the human (animal) by way of knowledge or, for that matter, through its opposite, ignorance, both of which continue to inscribe human existence into the binary operations of knowledge, functioning as the engine of the anthropological machine. Differently put, the human-animal relation is no longer "worked" in terms of knowledge, whether this working yields positive results—that is, renders the relation comprehensible—or fails and produces negative results, for instance, the conclusion that the human-animal relation remains unknowable. What Agamben proposes is that this vital relation does not, in fact, concern knowledge or, looking at it from the other end, as it were, that knowledge does not pertain to the human-animal duality. And this is the case because the proper mode of the human-animal relation is precisely one of nonknowledge. Notwithstanding all these qualifications, even as a zone of nonknowledge, this inoperative or inactive relation continues to figure the no longer simply human humanity but now does so precisely through nonknowledge (the inactivity of knowledge), which emerges in the aftermath of the *désoeuvrement* of the anthropological machine. One could say finally that what "constitutes" the no longer simply human (animal) described by Agamben is precisely the nonworking of the human relation to its "own" animality.

Now it is nonknowledge or aknowledge (91) that describes the "new" human-animal relation and does so specifically through the prism of inactivity. While the human, the animal, and thus their relation—now a relation of nonknowledge—all change, the schema that maps the "human" specifically *through* the human-animal relation remains, as (in)operative. Though its character changes from knowledge to nonknowledge, it remains in force as the "aknowable" figure of the human-animal relation. It is in this specific sense of reciprocally orienting the human and the animal that Agamben's figure of aknowledge remains vestigially humanist and anthropological. Clearly, it is not humanism in any historical or simple sense anymore, and it would be perhaps more appropriate to call it *nonhumanism*, since, if humanism is a domain of knowledge, then nonknowledge would be indicative of nonhumanism.

In his analysis, Agamben seems to identify humanism with the operations of the anthropological machine of knowledge, mastery, culture, technology, and so on. Therefore, when this machine is rendered inoperative, freeing the human-animal relation into the zone of nonknowledge, this release becomes effectively a reprieve from humanism. In Heidegger's analysis, though, humanism appears to have a broader scope, which is outlined precisely by the mapping of humanity in relation to animality, which renders the human-animal relation—whether conceived as knowledge or, as in Agamben, as nonknowledge—into the focal preoccupation of life. The very orientation of the human in terms of the animal constitutes a mark of the metaphysical, and thus humanist, revealing of the "human" way of being. From this Heideggerian perspective, it would not be enough, therefore, to release the human-animal relation into the zone of nonknowledge and to let it be beyond the operations of comprehension and the business of techno-calculations of modern life or of biopower. A further step, and perhaps ultimately a step in another direction, would be requisite—a letting go of the very figure that frames Agamben's reflection: the human-animal figure as the zone of aknowledge. This gesture would amount to releasing the human from its orientation toward animality, an orientation that remains originative of humanism. Another vector would become possible for the no longer simply human, namely, a decisive (re)orientation toward being, where being would be understood not as unconcealment or mastery of concealment, as it seems to be the case in Agamben, but instead as the event. The "human" would be revealed through the relation to the "there," to the site of this event, that is, in relation to its *Da*, and would thus be seen beyond the optics of nonknowledge and the inoperative human-animal relation. What is needed, however, for such a change of optics is a reading of *Da-sein* more attentive to Heidegger's thinking than the one Agamben presents, as Agamben's interpretation of *Dasein* in terms of an awakening from animal stupor all but loses Heidegger's critical insight into metaphysics and, more important here, his attempt to radically shift the thinking about the human away from the human-animal relation and toward *Da-sein*.

Toward the end of *The Open*, Agamben suggests that, from Heidegger's perspective, there are two possible scenarios with regard to humanism: "a) posthistorical man no longer preserves his own animality as undisclosable, but rather seeks to take it on and govern it by means of technology;

b) man, the shepherd of being, appropriates his own concealedness, his own animality, which neither remains hidden nor is made an object of mastery, but is thought as such, as pure abandonment" (80). The alternative that emerges here is between the technological mastery of animality and the human openness to concealedness as the hiatus, within the human, between the human and the animal. Agamben's conclusions are motivated by a particular interpretation of *Dasein*'s relation to animality in terms of anthropogenesis, that is, as the emergence of the human from the animal, as the "*becoming*-Dasein of living man" (68). Approaching the difference between the human and the animal genealogically, Agamben suggests that the *Da* is nothing other than the suspension of the captivation characteristic of animal behavior: "The jewel at the center of the human world and its *Lichtung* [clearing] is nothing but animal captivation; the wonder 'that beings *are*' is nothing but the grasping of the 'essential disruption' that occurs in the living being from its being exposed in a nonrevelation" (68).

In Agamben's scenario, the human emerges from the animal through the suspension of animal captivation, which opens the clearing as world. This anthropogenetic interpretation of the relation between *Dasein* and animality becomes the linchpin of Agamben's explanation of the play of concealedness and unconcealedness critical to *Dasein*: "This secret of unconcealedness must be unraveled in this sense; the *lēthē* that holds sway at the center of *alētheia*—the nontruth that also belongs originarily to the truth—is undisconcealedness, the not-open of the animal. The irresolvable struggle between unconcealedness and concealedness, between disconcealment and concealment, which defines the human world, is the internal struggle between man and animal" (69). This no longer Heideggerian version of the play of (un)concealment leads to the call to "restore to the closed, to the earth, and to *lēthē* their proper name of 'animal' and 'simply living being'" (73). At this point, Agamben's rather surprising conclusion becomes unavoidable: "*Dasein* is simply an animal that has learned to become bored; it has awakened *from* its own captivation *to* its own captivation" (70). The difference between animality and humanity lies thus in a type of captivation: while the animal is captivated by its environment, the human, having awakened from animal captivation, is, in turn, captivated by being. *Dasein* refers then to the genesis of the human out of the animal and into its specific zone of captivation by the question of being. Agamben transforms here the conflict between unconcealedness and concealedness into the conflict between the humanity of "man" and the animality of "man," a conflict that plays out within the genetically conceived humanist perspective.

For Agamben, *Dasein* is an animal that has awakened from being taken with its environment and, as a result, begins to try and master the animality from which it has awakened and of which it no longer has a "direct" experience. It is on this point that the difference between Agamben and Heidegger could not be sharper, since Heidegger makes it abundantly clear that *Dasein* cannot be explained on the basis of or in relation to animality. In fact, *Dasein* is precisely an attempt to think the human, and more precisely, the human way of being, apart from the originary link to animality, which Agamben forcefully restores in his anthropogenesis. What in fact motivates Heidegger's repeated questioning intrinsic to his thought is the need to refrain from starting the consideration of *Dasein in and as relation to* animality. This does not mean that the human does not have a body or an animal aspect to it, but that the understanding of the human should not depart from or base itself on its relation to animality. According to Heidegger, "Thus even what we attribute to man as animalitas on the basis of the comparison with 'beasts' is itself grounded in the essence of ek-sistence."[2] This is at least in part motivated by the difficulty of thinking living beings: "Of all the beings that are, presumably the most difficult to think about are living creatures, because on the one hand they are in a certain way most closely akin to us, and on the other are at the same time separated from our ek-sistent essence by an abyss" (230). Since Heidegger argues in *The Fundamental Concepts of Metaphysics* that from the perspective of being, or more properly, of the manner of being, the difference between the human and the animal is radical, it would not make sense, in a nonmetaphysical perspective, to speak any longer of a human animal (*homo animalis*) or a rational animal (*animal rationale*).[3] Though both are evidently beings, and seen biologically, animal beings, when thought ontologically, humanity and animality are of a different order, as it were, since their modes of being have entirely different parameters and require different types of analysis. What marks this difference is the fact that animals for Heidegger do not exist in terms of possibilities: animals are in fact closed *to* possibilities, as possibilities are never at issue for them, at least not qua possibilities. This is why the animal does not participate in opening a world by relating itself to possibilities and thus by making them possible, which means that the very terms of concealedness and unconcealedness, which pertain precisely to possibility, would be closed to animality. Since possibilities remain closed to animals, the concealedness at play in the possibilities opening up to *Dasein* in its temporal projection toward the future cannot be, within the horizon of Heidegger's thought, equated with the concealedness of animality *within* and *to* the human, as Agamben proposes. To put it simply, the concealedness characteristic of animality would be of a different order than the concealedness at play in *alētheia*, since the difference of animality would lie precisely in its being closed specifically to the very play of concealment and unconcealment. While Agamben understands the genesis of the human as the emergence of the difference, and at the same time the passage, within the human between the animal and the human, ontologically, there can be no such passage for Heidegger.

Dasein for Heidegger cannot, therefore, indicate the "between" or the caesura of the animal and the human, which defines the nonknowledge of the human in the aftermath of the anthropological machine in Agamben. This is principally the case because *Da-sein* refers "less" to humans than to the site of the crossing between the elements of what Heidegger in his later writings describes as the "worlding" of the world. Indicating the relation to being brought into the open and kept in play through the human being, the hyphen in *Da-sein* enacts the crossing of the fourfold, in which the animal, the human, and the suspension of the passage between them can come into play to begin with. In Agamben, the posthumanist "human" designates the suspension of the animal/human duality, whereas in Heidegger, *Da-sein* marks effectively a sidestepping and a regrounding of the humanist doublet. Rather than the suspended between of the double being (human/animal), *Da-sein* pertains to a different order of relations than the ones between humanity and animality. It does not suspend the passage between them but instead shifts the entire terrain for the consideration of "human" being, that is, of the way of being that is human, displacing the focus from animality to being and the saying of its event. By contrast, Agamben does not go beyond the anthropological machine, but, in a way, prefers to unwork it, so that the animating link between the animal and the human that has energized humanism becomes inoperative.

This underscoring of the hiatus and the silence between the animal and the human suggests nonetheless that the perspective of Agamben's analysis, its "grounding" in the nonknowledge between the human and the animal, remains unchanged. While Agamben cuts and suspends the human-animal passage, Heidegger attempts to move not only beyond the horizon of humanism and anthropocentrism, but also beyond the human/animal doublet understood as the decisive, even if "aknowable," relation for humanity. Perhaps the clearest indication of this difference between Agamben and Heidegger is the discourse of "life" that underpins *The Open*, to such a degree that Agamben does not hesitate to characterize *Dasein*, insofar as it corresponds to animality awakened from its captivation, as a figure of life. The concept of life appears to be the operative way of describing "being" in *The Open*, as, for instance, when the *désoeuvrement* of the human and the animal becomes "the supreme and unsavable figure of life" (87), or when, returning to the illustration of the banquet of the righteous at the end of *The Open*, Agamben speaks about the portrayed humans/animals as "living beings" (92). By contrast, Heidegger decisively refuses to inscribe *Dasein* and the human into the horizon of living beings: "All anthropology continues to be dominated by the idea that man is an organism [*Lebewesen*; literally, "living being"]. . . . If we are to think of man not as an organism but a *human* being [*Menschenwesen*], we must first give attention to the fact that man is that being who has his being by pointing to what is. . . ."[4] In a similar vein, in "Letter on Humanism," Heidegger remarks, "The human body is something essentially other than an animal organism" (228).

There is no room here to spell out the reasons for or the implications of Heidegger's decisive critique of the notion of a living being. Suffice it to note that the framework of the living being maintained in *The Open* may be the linchpin of the difference between Agamben's and Heidegger's approaches. Framing the discussion of the human in terms of living beings continues the metaphysical and thus anthropological elaboration of *humanitas*. As I suggested earlier, though inoperative, the human-animal relation remains "vital" in Agamben and does so in a crucially double sense: it is not only "alive" and crucial in Agamben's thinking but pertains precisely to the problematic of "life," the problematic that holds the animal and the human together in *The Open*. Against this way of thinking, Heidegger repeatedly insists on the necessity of initiating the thinking of the human not in relation to life, that is, as a living being, but instead in relation to *Da-sein*. The relation to the *Da*, to the site of the event, takes priority over the human-animal relation, which in fact can be thought only within the opening of the *Da* and its characteristic mode of being.

Obviously, human beings have a relation to animality and can be understood as "animal" in terms of their "biological" bodies, yet neither humanity nor human corporeality can be defined on the basis of animality or animal corporeality. Because of this continuously receding difference, which, strictly speaking, could not even be conceived of as difference, Heidegger proposes to think humanity, or at least to begin to think humanity, without an initial relation to animality. Along the same lines, it is not the human body that, for Heidegger, determines the human "understanding" and relation to being, but, conversely, it is the human relation to being, understood as a response to being's call or claim, that shapes human corporeality as unfolding in terms of the play of futurity and historicity. Animal corporeality is shaped, by contrast, through captivation (*Benommenheit*). From the point of view of biological sciences, animal and human bodies can be seen along the same spectrum as variants of an essentially animal, *living* body. For Heidegger, this represents, however, a way of understanding the body and corporeality as being decisively rooted in the metaphysical conceptualization of matter, body, and corporeality, in parallel with their supposed opposites: spirit, soul, and mind. Though Heidegger does not develop this reflection beyond some indicative remarks on spatiality in *Being and Time*, human corporeality would have to be thought *first* from within and in its relation to the "there" of *Da-sein* and not in terms of the body, that is, not as an animal or a human-animal body to begin with.[5] To put it simply, one would have to consider "human" corporeality neither in terms of its "humanity" nor in terms of its "animality," and thus certainly not in terms of life, but through its relation to and its participation in the historically unfolding *Da*/there of being. To that extent, neither animality nor humanity, considered as human animality, enters into this initial level of the consideration of the spatio-temporality of the *Da*. Human being— that is, the mode of being given to humans—is not related, to begin with, to animality, but instead to being, or more precisely, to the question of being, that is, to the fact that being is always and already *in question* for the way that humans are, namely, as being-there, as *Da-sein*. This way of beginning with the *Da* rather than with humanity or animality permeates all of Heidegger's work, even if it is not always explicit in *Being and Time*. Therefore, seeing *Dasein* in terms of its genesis from animality, as an awakening from and, in effect, an altered mode of animality, seems to foreclose the line of questioning opened up by Heidegger in his consideration of *Dasein*. It means, in a way, to cover *Dasein* over and thus to risk a crucial misunderstanding of the orientation of Heidegger's rethinking of the human.

"Letter on Humanism" could not be more explicit in underscoring this critical shift from the notion of a living being to what Heidegger calls *ek-*

sistence as the new optics for thinking the human, where *ek-sistence* refers precisely to "standing in the clearing of Being" (228), that is, in the *Da* of *Da-sein*.

> "The 'substance' of man is ek-sistence" says nothing else but that the way that man in his proper essence becomes present to Being is ecstatic inherence in the truth of Being. Through this determination of the essence of man the humanistic interpretations of man as *animal rationale*, as "person," as spiritual-ensouled-bodily being, are not declared false and thrust aside. Rather, the sole implication is that the highest determinations of the essence of man in humanism still do not realize the proper dignity [*Würde*] of man. To that extent the thinking in *Being and Time* is against humanism. But this opposition does not mean that such thinking aligns itself against the humane and advocates the inhuman, that it promotes the inhumane and deprecates the dignity of man. Humanism is opposed because it does not set the *humanitas* of man high enough. (233–34)[6]

Heidegger remarks here that "the *humanitas* of man" is not thought "high enough" precisely in order to forestall the misunderstanding of his opposition to humanism as an endorsement of inhumanity or inhumanness. As he states emphatically on several occasions, a critique of humanism does not mean "a defense of the inhuman and a glorification of barbaric brutality" (249). Quite the opposite is at stake, as "opposition to 'humanism' in no way implies a defense of the inhuman but rather opens other vistas" (250).[7] What is especially interesting is that it is not simply animality that Heidegger has in mind here when he mentions the fact that *humanitas* is not thought "high enough," but primarily reason, spirit, culture, values, and technology, as "Letter on Humanism" makes clear. This is the case because these metaphysical articulations of the human fail to think the dwelling characteristic of being human. The distinctive "dignity" of the human being comes from its participation in the event of being, which it helps bring into words and shelter. What Heidegger calls here the "dignity" (*Würde*) of the human does not lie first in rationality and the technological culture it has produced, or in spirituality and the culture of values it has engendered. It is "higher" than these human values in a very specific sense: it points beyond the human to the dignity and worthiness of being.

The other vista Heidegger has in mind here is the approach to the human through *ek-sistence* and *Da-sein*, where it is clear that the *Da* is not "first and foremost" about the body, animality, or life, on the one hand, and reason, spirit, or culture, on the other. In other words, the *Da* in *Da-sein* is not anthropic, or at least it is not initially determined in anthropic terms. Rather, the anthropic and life-oriented functions and preoccupations of the human, from the bodily to the spiritual, unfold always already from within the *Da*, which marks the originary "human" relation to being, a relation that tunes and disposes (*stimmt*) both corporeality and rationality. Because of this disposing that makes determination (*bestimmen*) possible to begin with, the *Da* and its vectors of attuning the human remain "prior," as it were, to the mind-body relation. This is why in its relation to *Da-sein* the human is not considered for itself, or for its sake, as it were, but instead for the sake of being. What takes priority here is in a way not the human in relation to the *Da*, but the *Da* itself, specifically insofar as the *Da* opens up in relation to the human. This is the case because, beginning probably with *Contributions to Philosophy*, the *Da* in *Da-sein* both is and is not "human," strictly speaking.[8] The *Da* is the site of being's always already beginning as the event, but this event can take place only with the human participation in the *Da* and, more precisely, with the human being displaced into and as the "there" of being, as *Da-sein*. Differently put, while not human, the *Da* does not open up without the human being "there."

The human is then very much under consideration for Heidegger, but not "as such" and only to the extent of its relation to being, that is, to the extent to which the access to the question of being is through *Da-sein* and thus through the human participation in the *Da*. What interests Heidegger is being's unfolding as the event; yet the event and its beginning pivot precisely on the displacement of the human being into and as the *Da* of *Da-sein*. What is at issue for Agamben is the anthropological machine of humanism, the human, and the animal. For Heidegger, however, it is neither the animal nor the human but being, for, as he announces in *Being and Time*, it is the question of being that he intends to explore and not one of humanity.[9] It is for this reason that the "strange" humanism Heidegger proposes in "Letter on Humanism" is less about thinking the human-animal than nearness to being: "It is a humanism that thinks humanity of man from nearness to Being. And not nearness to and separation from the animal. But at the same time it is a humanism in which not man but man's historical essence is at stake in its provenance from the truth of Being" (245). This nearness to being as the event becomes possible through a nonmetaphysical and thus "nonhuman" being in the there of *Da-sein*. In the end, all of Heidegger's reflections on humanity and animality serve to elaborate ways of thinking nearness to being nonmetaphysically.

Perhaps the most important aspect of this rethinking of the human in terms of *Da-sein* is that it is not life or animality but the human relation to being and to its clearing (*Lichtung*), that is, to its *Da*, and its catalyzing role in the unfolding of being as the event, that is the originative momentum of the being given to humans. This moment is, however, not "human" as such, without at the same time being animal, technological, inhuman, and so on, since terms like *animal, inhuman,* and *technological* become operative always already in relation to the human. To think the nonhuman or post-human amounts to thinking humanity still in the humanist and anthropological cast, by directing the inquiry toward the other of the human, whether construed as in-, non-, or posthuman. Heidegger's main interest lies instead in engaging with being, and not with the human being, even though it becomes possible to ask the question of being only in terms of the relation of being to humans. More specifically, it is a matter of thinking the relation between being and what in the human mode of being renders it receptive, as it were, to being's claim, that is, to what enables being to be at issue for humans. It is specifically this relation between being and the way it gives its *Da*, the there of its clearing, to be experienced by humans that preoccupies Heidegger above all else. In fact, the human being needs to be thought from this relation between being and the *Da*, and only in this perspective can one subsequently address the relation between the human mode of being and the animal mode of being. Interrupting the anthropological machine of humanism will not, therefore, address by itself the way in which being's relation to its "there" is originative of the essential unfolding (*Wesen*) of the human.

Heidegger's critical point is that what unfolds essentially (*Wesen*) in humans is not human to begin with, because it is the *Da*, the clearing of being, to which humans belong and in relation to which they come to be "human" in the first place. In his recently published writings from the period around World War II, in particular in *Metaphysik und Nihilismus* and *Nietzsche Seminaren 1937 und 1944*,[10] Heidegger adopts an even more decisively "nonhuman" and nonmetaphysical terminology oriented toward being, nihilation, and the event (*das Ereignis*). It is worth recalling here that in the margin of his copy of "Letter," Heidegger handwrote a remark about the "Letter" consciously speaking the language of metaphysics.[11] What accounts for this metaphysical language is, on the one hand, the unavailability in 1946 of several texts Heidegger had written between 1936 and the appearance of the "Letter" a decade later, and on the other, the need, so clear in the text itself, to forestall the misunderstanding of the critique of humanism for antihumanism or, worse, for inhumanity. The texts written during the years preceding the "Letter" are more explicit and forceful about the necessity to oppose the pervasive *Vermenschung*—anthropization or humanization—of being. Two moments in the recently published *Metaphysik und Nihilismus* in particular are important and even poignant here, as they shift the emphasis from the human being to the event. "That *Being and Time* aims to do away with man and his priority in philosophy, and give its due simply to the happening of being, that with *Da-sein* it is not only the subjectivity of man but the role of man that is shattered, will be one day realized."[12] Nonmetaphysical thinking requires putting aside (*beseitigen*) not just subjectivity but effectively "man," and thus also humanism. As Heidegger indicates, thinking needs to give its due, literally, its worth (*Würde*), to being and cease originating its questioning with "man" and centering it either directly on the human or in relation to humans. This appears to be a rather audacious move, especially in the context of the inhumanity and barbarity of the war taking place at the time and given Heidegger's own engagement with National Socialism in the 1930s. Yet when we consider this move carefully, we realize that it is forced precisely by the need to radically reclaim the worth and dignity (*Würde*) proper to the human and by Heidegger's conviction that this can be done only by recasting the human beyond and apart from any humanist/metaphysical optics.

Humanism works on the assumption that in order to accord proper dignity and significance to humans, the human being needs to be placed in the center, as it were, and understood as the most important aspect of being, thus becoming the aim of humanity itself. As a result, all that exists comes to be oriented toward and around humans, and thus submitted to their evaluation: it comes to exist in terms of its value as determined in relation to the human. For Heidegger, however, to restore the "proper" dignity to the human, the human must be (dis)placed in relation to being. To restore *humanitas* to humans, it is necessary—however paradoxical it might seem at first—to abandon the humanistic, and metaphysical, conceptualization of being (primarily as "life") and to reclaim instead the dignity, the worth (*Würde*) of being.

Yet what is this dignity and worth(iness) of being? It is clearly not a value, not something that could be comprehended, grasped, and articulated as an idea or (re)presented as a value. The worth of being lies in its specific temporal-historical force of nihilation, which enables possibilities and opens being as the futurally unfolding nexus of relations, constitutive of the historical, ceaselessly reenacted world. In short, the "dignity" of being is its unfolding as the event, an event in which the metaphysical determinations of being are called into question in such a way that being can be released from them and let be as the event with its nihilating force. The dignity of being lies thus in being's release into the sway (*Walten*) of the nihilating force of the event—the force that, in the same gesture, empties all that is of "essence" and opens it up to the historical, spatio-temporal span of possibilities. Such an event is, therefore, always double: on the one hand, it undermines the entire history of various metaphysical manifestations of being, from ancient Greece to the modern "essence" of technology, and, on the other, this undermining happens each time as an always singular displacement into the openness of the there of *Da-sein* as the site where the event can be experienced as event. This double span of the event, both always historical and each time singular, constitutes the worth(iness) of being. It is thus possible to say that the dignity of being lies in the each time singular force with which its event is "historial"; that is, it nihilates the historical sedimentations of being, freeing future possibilities. The dignity of being is the each time singular possibility of the freedom of its event, its release from the historically determining metaphysical frameworks, that is, its "freedom" to be historical in a singular way.

It is in this perspective of the worth of being as the force of nihilation, the force that pervades and orients the *Da* of *Da-sein*, that one can glimpse the dignity and worth of the human. Because of their participation in the *Da*, humans' dignity lies in their capacity to be attentive, or "mindful"—in the sense of *Besinnung*—of this originative force of nihilation and of its dignity, its worth, that is, of its pervading importance, for all being.[13] Obviously dignity here is thought in a particular, "nonmetaphysical" sense, which pertains to the way in which nihilation gives everything to be as what it is.

It would require a separate occasion to explain why Heidegger chooses the term *Würde* to refer to being and the event. On the one hand, the term harks back to Kant and his way of describing the "dignity" of "man" and would underline here a critical displacement of focus from "man" to "being." More important, however, the term differentiates between a nonmetaphysical thinking about being and the human in terms of *Würde* and the metaphysical estimation of beings in terms of value (*Wert*). Another remark from "Letter on Humanism" makes this quite clear:

> To think against "values" is not to maintain that everything interpreted as "a value"—"culture," "art," "science," "human dignity" [*Menschenwürde*], "world," and "God"—is valueless [*wertlos*]. Rather, it is important finally to realize that precisely through the characterization of something as "a value" [*Wert*] what is so valued [*Gewertete*] is robbed of its worth [*seiner Würde beraubt wird*]. That is to say, by the assessment of something as a value what is valued is admitted only as an object of man's estimation. (251)[14]

Humanism and metaphysics estimate any and everything in relation to the human and thus constitute it as "a value." For Heidegger, thinking in terms of value is a distinctive form of *Vermenschung*, or of humanization or anthropization. To think against values, as Heidegger repeatedly underscores it, does not mean to automatically assert that beings are without value or importance, but instead to shift the perspective in which beings come to be seen to begin with. Value is established in relation to the human, and as such it "robs" what exists of its worth (*Würde*). In distinction from "value," "worth" becomes disclosed first and foremost *in relation to being*, and *not humans*, that is, it unfolds primarily in terms of the force of nihilation and possibility, and never comes to stand as "a value," whether ontological, representational, epistemological, or ethical. Value is measured in the currency of lastingness, graspability, possession, standing presence, and so on. Worth, on the other hand, is (dis)measured in terms of maintaining possibility and nihilation in force. In a way, worth is what "nihilates" values and makes reevaluation possible, say, in the manner in which Nietzsche attempts a revalorization of all values. To the extent to which value thinking, by making what exists into a value, renders being graspable and estimable, at the same time it "robs" what exists of its worth, that is, of its "dignity" of *being*, of being temporal, and of existing in terms of nihilation and the enabling of possibilities. To the degree to which humanist approaches think the human in terms of being and thus also of values, they never reach or engage with the worth "proper" to the human way of being, namely, with its role of sheltering the force of nihilation.

It is this tension between the humanist value of the human (as rational animal) and its "nonmetaphysical" dignity of being—that is, the experience of mortality and nihilation, which is, at the same time, the experience of futurity, enabling, and possibility—that leads Heidegger to "oppose" humanism. The value of the human being is its *humanitas*, whether measured in terms of rationality or animality, soul or body, and so on, but its worth/dignity is mortality, because it is only as mortals that humans are *both* open to nihilation *and* capable of maintaining and sparing its enabling force. The dignity and worth of the human mode of being is "measured," as it were, by "letting be," that is, by the ability to let nihilation nihilate, and thus to enable possibilities without their becoming arrested into a system of ideas, values, possessions, and so on. This is precisely why, in order to restore dignity/worthiness to the human, human being must be dis-valued, that is, freed from being (mis)understood, determined, or experienced first and foremost in terms of value.

This dignity, or the worth, of the human being is not, strictly speaking, "human," as it refers to the attentiveness and "care" that, in its being, a human being can give to the clearing of being itself. The *ek-sistence* of humans concerns above all the essential unfolding (*Wesen*) of being, to which humans belong and in which they take part. Another remark about the "essence" (*Wesen*) of the human from *Metaphysik und Nihilismus* makes this point explicitly: "Dieses Wesen is nicht Menschliches . . ." (This "essence" is not something human) (240). It is not of the human, or it is, in other words, nothing human. What "unfolds essentially" in humans, namely, the "there"/*Da*, is not human, at least not in any humanist sense of the term. In fact in this specific sense, *Da-sein* is no longer either human or animal but instead takes its shape as the site of the event. While the there—as the site where being comes into the open (into the *Da*) and unconceals itself—is not human, it cannot stay open without human participation. The dignity or worth of the human (way of) being lies in this capacity to take part *in* and *as* the *Da*, by bringing into words the event and the unconcealment of being. The distinctiveness of this (human) way of being lies thus in the ability to recognize the "prior" dignity and worth of being, the worthiness from which the character, the essence, and the dignity of humans unfold. What renders humans unlike any other being is their "capacity" to recognize and shelter—both in thinking and in acting—the nihilating and enabling force of being and keep it as critical of value-oriented thinking. Thus, their "own" importance comes only and always already *in relation to* and *by way of* the dignity (*Würde*) of being: not from themselves, not in terms of their "self" or self-relation. Therefore, in order to think the worth and the *humanitas* of humans "high enough," that is, "properly" in terms of how they are given to be, one has to disanthropize and "dishumanize" thinking, to oppose the *Vermenschung* with an *Entmenschung*. The dignity of humans comes from preserving, sparing, and sheltering what is genuinely or originarily—that is, first and foremost—worthy and most "dignified," namely, *being*, understood as the worlding of world. To experience their *humanitas*, humans have to recognize their role of "assistance" to being, as is clearly reflected in Heidegger's manner of writing in "Letter on Humanism," where the emphasis repeatedly falls on how *being relates* to *ek-sistence* and not on how *the human (subject) relates* to being (235). Heidegger thus reverses both the question—no longer of the human but of being—and the grammar in which it is articulated in order to try and begin thinking the question of being from the event.

This conceptual and grammatical reversal underscores the fact that the human is not the lord of being, with the implication that she or he is likewise not a slave of being, obedient to its dictates. As the phrase "the shepherd of being" implies,[15] human "dignity" lies in tending to being rather than in either mastering or, conversely, obeying it. Though not being the master of being appears to lessen the human, Heidegger points out that this apparent "less" is actually a gain: "Man loses nothing in this 'less'; rather he gains in that he attains the truth of Being" (245). By releasing being from the master/slave dialectic, the humans also free themselves, dis-covering their "worth" of mortals, who, precisely by being mortal, that is, by *ek-sisting* in a temporal and finite manner, are capable of experiencing and bringing into words the "dignity" of the nihilating force that "clears" the *Da*, the site of being, opening it ceaselessly toward the future and the arising possibilities.

In this perspective, one can return to and read the hiatus in Agamben as a form of letting the difference between humanity and animality be, precisely without trying to grasp or master it. In thus leaving the difference unworked and inoperative, what comes to the fore is perhaps the "essence" of the human, defined precisely as "letting be." Though this moment is not entirely clear or fleshed out in Agamben's text, it appears that leaving the difference unworked becomes a matter of a human decision and act, in which the human acts in order not to act (not to work the difference at the core of the anthropological machine). "Letting be" would become the defining optics of being human, one in which the idea of the human constructed as active and acting, as doing and working, and thus as grasping, knowing, and mastering—working and mastering first and foremost the difference between the human and the animal—would become possible to begin with. To a certain extent, doing and working would then be a way both of "accomplishing" the human as a rational animal and simultaneously of obscuring its originary manner of letting be. This way of reading the hiatus would indeed bring Agamben very close to Heidegger's idea of the human as illuminated primarily in terms of the response to being and as oriented by letting be. In Heidegger, this letting be makes room precisely for the worth (*Würde*) of being, that is, of nihilation and enabling, without closing off and apportioning being into comprehensible and graspable values, which effectively conceal the temporal and finite momentum of being. The radicalism of the Heideggerian position lies in granting this "primary" worth not to human-animal life but to being's momentum of enabling and rendering possible through nihilation. This radicalism underscores the originative force of the nothing—of the event of being— whose nihilating momentum and enabling force are perhaps "more" worthy, that is, more question-worthy, because more properly *of* being, than the human. It is at this point that Heidegger's thought appears capable of "leaving aside" not just philosophical metaphysics but the Western perspective itself.

Notes

1. Giorgio Agamben, *The Open: Man and Animal*, trans. Kevin Attell (Stanford, CA: Stanford University Press, 2004), 1. Hereafter cited parenthetically by page number.
2. Martin Heidegger, "Letter on Humanism," in *Martin Heidegger: Basic Writings*, trans. and ed. David Farrell Krell (1977; New York: HarperCollins, 1993), 228. Hereafter cited parenthetically by page number.
3. Martin Heidegger, *The Fundamental Concepts of Metaphysics: World, Finitude, Solitude*, trans. William McNeill and Nicholas Walker (Bloomington: Indiana University Press, 1995).
4. Martin Heidegger, *What Is Called Thinking?* trans. J. Glenn Gray (New York: Harper and Row, 1968), 148–49.
5. Martin Heidegger, *Being and Time*, trans. John Macquarrie and Edward Robinson (New York: Harper and Row, 1962).
6. Martin Heidegger, "Brief über den Humanismus" ("Letter on Humanism"), in *Wegmarken* (*Pathmarks*) (Frankfurt am Main: Vittorio Klostermann, 1978), 327. My translation.
7. For an excellent and extended discussion of Heidegger's rethinking of *humanitas* in "Letter on Humanism," see Françoise Dastur, *Heidegger et la question anthropologique* (Leuven: Peeters, 2003), 65–82.
8. Martin Heidegger, *Contributions to Philosophy (from Enowning)*, trans. Parvis Emad and Kenneth Maly (Bloomington: Indiana University Press, 1999).
9. Heidegger, *Being and Time*, 37.

10 Martin Heidegger, *Metaphysik und Nihilismus*, *Gesamtausgabe*, vol. 67 (Frankfurt am Main: Vittorio Klostermann, 1999), and *Nietzsche: Seminaren 1937 und 1944*, *Gesamtausgabe*, vol. 87 (Frankfurt am Main: Vittorio Klostermann, 2004).
11 This is mentioned by Jean-François Mattei in "L'homme dans le Quadriparti" ("The Man in the Fourfold"), in *Heidegger et la question de l'humanisme: Faits, concepts, débats*, ed. Bruno Pinchard (Paris: Presses Universitaires de France, 2005), 257.
12 Heidegger, *Metaphysik und Nihilismus*, 90. My translation.
13 For an extended discussion of nihilation and its relation to thinking politics, see my essay "The Other Politics: Anthropocentrism, Power, Nihilation," in *Letting Be: Fred Dallmayr's Cosmopolitical Vision*, ed. Stephen F. Schneck (Notre Dame, IN: University of Notre Dame Press, 2006).
14 See also Heidegger, "Brief über den Humanismus," 345. My translation.
15 In a different context, it would be necessary to engage more critically with the phrase "the shepherd of being." However, here I refer to it to simply to note that Heidegger's emphasis falls deliberately on human tending and attentiveness to being as distinct from the "active" attitude of production, manipulation, or mastery.

Notes on Contributors

ANDREW BENJAMIN is a professor of critical theory and philosophical aesthetics in the Centre for Comparative Literature and Cultural Studies at Monash University, Melbourne. His most recent book is *Style and Time: Essays on the Politics of Appearance* (Northwestern University Press, 2006).

CLAIRE COLEBROOK teaches English literature at the University of Edinburgh. She is the author of *New Literary Histories: New Historicism and Contemporary Criticism* (Manchester University Press, 1997), *Ethics and Representation: From Kant to Post-Structuralism* (Edinburgh University Press, 2000), *Gilles Deleuze* (Routledge, 2005), *Irony in the Work of Philosophy* (University of Nebraska Press, 2003), *Understanding Deleuze* (Allen and Unwin, 2003), *Irony: The New Critical Idiom* (Routledge, 2003), *Gender* (Palgrave Macmillan, 2005), and *Deleuze: A Guide for the Perplexed* (Continuum International, 2006). Her most recent book is *Milton: Evil and Literary History* (Continuum, 2007).

JEAN-PHILIPPE DERANTY teaches philosophy at Macquarie University, Sydney. He is completing a manuscript on Axel Honneth's theory of recognition, to be published by Brill in 2008.

PENELOPE DEUTSCHER is an associate professor of philosophy at Northwestern University. She is the author of *Yielding Gender: Feminism, Deconstruction, and the History of Philosophy* (Routledge, 1997), *A Politics of Impossible Difference: The Later Work of Luce Irigaray* (Cornell University Press, 2002), *How to Read Derrida* (Granta, 2005), and *The Philosophy of Simone de Beauvoir: Conversion, Ambiguity, Resistance* (Cambridge University Press, forthcoming).

ELEANOR KAUFMAN is an associate professor of comparative literature and French and Francophone studies at the University of California, Los Angeles. She is the coeditor of *Deleuze and Guattari: New Mappings in Politics, Philosophy, and Culture* (University of Minnesota Press, 1998) and the author of *The Delirium of Praise: Bataille, Blanchot, Deleuze, Foucault, Klossowski* (Johns Hopkins University Press, 2001) and *At Odds with Badiou: Politics, Dialectics, and Religion from Sartre and Deleuze to Lacan and Agamben* (Columbia University Press, forthcoming).

ADRIAN MACKENZIE researches in the area of technology, science, and culture at the Centre for Social and Economic Aspects of Genomics, Lancaster University. He has published two books on technology, *Transductions: Bodies and Machines at Speed* (Continuum, 2002) and *Cutting Code: Software and Sociality* (Peter Lang, 2006), and articles on media, science, and culture.

CATHERINE MILLS is a lecturer in philosophy at the University of New South Wales in Sydney. Her research areas include contemporary continental philosophy, especially in relation to questions of ethics, politics, and law, and feminist philosophy. She has previously published in journals such as *differences*, *The Journal of Political Philosophy*, and *Australian Feminist Studies*, among others. She has contributed to several recent collections of essays on the work of Giorgio Agamben and is currently completing a manuscript titled *The Philosophy of Agamben* for Acumen Press.

ALISON ROSS teaches in the Centre for Comparative Literature and Cultural Studies at Monash University, Melbourne. She is the author of *The Aesthetic Paths of Philosophy: Presentation in Kant, Heidegger, Lacoue-Labarthe, and Nancy* (Stanford University Press, 2007).

LEE SPINKS is a senior lecturer in the Department of English Literature at the University of Edinburgh. He is the author of *Friedrich Nietzsche* (Routledge, 2003) and numerous articles on critical theory, contemporary American literature, and postcolonial literature. His study of Michael Ondaatje will appear with Manchester University Press in 2008.

EWA PŁONOWSKA ZIAREK is Julian Park Professor of Comparative Literature at the State University of New York at Buffalo. She is the author of *The Rhetoric of Failure: Deconstruction of Skepticism, Reinvention of Modernism* (State University of New York Press, 1995), *An Ethics of Dissensus: Feminism, Postmodernity, and the Politics of Radical Democracy* (Stanford University Press, 2001), and the coeditor of *Revolt, Affect, Collectivity: The Unstable Boundaries of Kristeva's Polis* (State University of New York Press, 2005) and *Intermedialities: Philosophy, Art, Politics* (Rowman and Littlefield, forthcoming).

KRZYSZTOF ZIAREK is a professor of comparative literature at the State University of New York at Buffalo. He is the author of *Inflected Language* (State University of New York Press, 1994), *The Historicity of Experience* (Northwestern University Press, 2001), and *The Force of Art* (Stanford University Press, 2004), and the coeditor of *Future Crossings* (Northwestern University Press, 2000) and *Adorno and Heidegger* (Stanford University Press, 2008). He is the author of two books of poetry in Polish, *Zaimejlowane z Polski* (Archeton Druk, 2000) and *Sąd dostateczny* (Klematis, 2005).

New from Stanford University Press

Psyche
Inventions of the Other, Volume II

JACQUES DERRIDA,
Edited by PEGGY KAMUF and
ELIZABETH G. ROTTENBERG

"This monumental collection of essays shows Derrida at his brilliant best, across a vast and diverse range of topics, texts, authors and manners. *Psyche* is among the richest and most diverse of all Derrida's books, and a testimony to the extraordinary depth and vigor of deconstructive thought."
—Geoffrey Bennington,
Emory University

Meridian: Crossing Aesthetics
Feb 2008 $24.95 paper $65.00 cloth

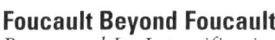

Foucault Beyond Foucault
Power and Its Intensifications since 1984

JEFFREY T. NEALON

"*Foucault Beyond Foucault* is the first major renovation of the critical representation of Foucault's system in the past twenty years. Jeffrey Nealon successfully challenges the critical prejudices and assumptions that have defined Foucault's legacy."
—Gregg Lambert,
Syracuse University

$21.95 paper $55.00 cloth

Selected Writings

SARAH KOFMAN,
Edited by THOMAS ALBRECHT
with GEORGIA ALBERT and
ELIZABETH G. ROTTENBERG
Introduction by
JACQUES DERRIDA

This book is a comprehensive anthology of significant essays and book excerpts by the postwar French philosopher and theorist Sarah Kofman (1934-1994).

Meridian: Crossing Aesthetics
$24.95 paper $65.00 cloth

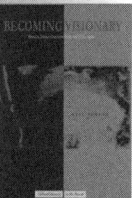

Becoming Visionary
Brian De Palma's Cinematic Education of the Senses

EYAL PERETZ, with a
Foreword by
STANLEY CAVELL

"Modestly framed as an attempt to lift Brian de Palma from the category of Hitchcock imitator and restore his unique vision, *Becoming Visionary* opens into a brilliant speculation on cinema itself, on cinema as an apparatus of sensation. Philosophy and film have never seemed more suited for mutual enlightenment than in Peretz's deft analysis."
—Joan Copjec,
SUNY Buffalo

Cultural Memory in the Present
$21.95 paper $55.00 cloth

Exemplarity and Chosenness
Rosenzweig and Derrida on the Nation of Philosophy

DANA HOLLANDER

"Dana Hollander's *Exemplarity and Chosenness* represents a major contribution to the study of Derrida and Rosenzweig and the rapidly expanding field of new religio-philosophical studies. This is a book that must be read by anyone interested in Jewish Studies, religion and philosophy, or critical theory today."
—Kenneth Reinhard,
University of California, Los Angeles

Cultural Memory in the Present
$60.00 cloth

University Press
800.621.2736   www.sup.org

new from Duke

An Empire of Indifference
American War and the Financial Logic of Risk Management
RANDY MARTIN
Social Text books   232 pages, paper, $21.95

Stages of Emergency
Cold War Nuclear Civil Defense
TRACY C. DAVIS
432 pages, 58 illustrations, paper, $24.95

Lenin Reloaded
Toward a Politics of Truth
sic vii
**SEBASTIAN BUDGEN, STATHIS KOUVELAKIS,
& SLAVOJ ŽIŽEK, EDITORS**
[sic] Series   352 pages, paper, $23.95

Imagining Our Americas
Toward a Transnational Frame
SANDHYA SHUKLA AND HEIDI TINSMAN, EDITORS
Radical Perspectives   424 pages, 12 illustrations, paper, $24.95

Chicana Art
The Politics of Spiritual and Aesthetic Altarities
LAURA E. PÉREZ
Objects/Histories   408 pages, 90 illustrations (incl. 73 in color), paper, $24.95

Duke University Press

toll-free 1–888–651–0122   www.dukepress.edu

Beyond Exoticism
Western Music and the World
TIMOTHY D. TAYLOR
Refiguring American Music   328 pages, 16 illustrations, paper, $22.95

Sessue Hayakawa
Silent Cinema and Transnational Stardom
DAISUKE MIYAO
A John Hope Franklin Center Book   400 pages, 23 illustrations, paper, $23.95

Wallowing in Sex
The New Sexual Culture of 1970s American Television
ELANA LEVINE
Console-ing Passions   336 pages, 29 b&w photos, paper, $22.95

Getting Loose
Lifestyle Consumption in the 1970s
SAM BINKLEY
336 pages, 27 illustrations, paper, $22.95

Interventions into Modernist Cultures
Poetry from Beyond the Empty Screen
AMIE ELIZABETH PARRY
Perverse Modernities   200 pages, paper, $22.95

The Enemy
RAFAEL CAMPO
112 pages, paper, $17.95

TOPIA
CANADIAN JOURNAL OF CULTURAL STUDIES

Published by:
Wilfrid Laurier University Press
in partnership with
Cape Breton University Press

Subscription address:
E-mail: press@press.wlu.ca
Wilfrid Laurier University Press
75 University Avenue West
Waterloo, Ontario
Canada N2L 3C5
Telephone: (519) 884-0710 Ext. # 6124
Fax: (519) 725-1399

Editorial address:
TOPIA
E-mail: topia@yorku.ca
Graduate Programme in
Communication & Culture
TEL Centre, Rm 3017, 88 The Pond Road
York University, 4700 Keele Street
Toronto, Ontario M3J 1P3
Telephone: (416) 736-2100 Ext. # 22238

Editor: Jody Berland
Annual Subscription Rates:
Institutional $80 Individual $40 Student $25
(Orders outside Canada remit payment in U.S. dollars)
ISSN 1206-0143

yorku.ca/topia